Phy Psychonauts
by Boaz Yemini

Why Science and Psychedelics Go Hand-In-Hand

Physics for Psychonauts
Why Science and Psychedelics Go Hand-In-Hand

Published by
Boaz Yemini

Tel-Aviv-Yafo, Israel
PhyForPsy@gmail.com
Telegram: @BoazQuant
Copyright © 2021 בועז ימיני
First published in 2022
All rights reserved

Art by Maksim Marinin

No part of this product may be reproduced in any manner without the prior written permission of the publisher.
ISBN: 9798405324784 (paperback)
Also available as Kindle
A disclaimer – All the ideas herein are believed to be accurate at the time of publishing, but, the author and publisher accept no liability for any type of errors, or for any actions, inspired by this book.

*For Mom & Dad,
who paid for my first trips.*

Table of Contents

Chapter 0 – Pre-Intro: 1

Foreword: 2 | Preface: 4 | Acknowledgements: 6

Chapter 1 – An Intro: 9

State of Things: 16 | This That You Read: 20 | Some Definitions: 23

Chapter 2 – Some History: 31

From 2000: 33 | 1950 to 99: 44 | 1900 to 49: 52

From Newton to the Electron: 64

Chapter 3 – On Models and Averages: 69

A Model: 71 | Wave Function: 76 | Hidden Variables: 80

The Case of $\sqrt{-1}$: 83

Chapter 4 – The Matter We Are: 87

Derivatives: 88 | Action at a Point: 91 || Entanglement(ing): 95

Chapter 5 – The Psychedelic Episode: 101

Seeing: 105 || Psilocybin: 109 || 5-MeO-DMT: 112 || Necessity: 117

Chapter 6 – The Matter That Matters: 123

An AetherFluid: 127 || Waves in/of Space: 130 || Knot at a Point: 133

Chapter 7 – Non-Matters: 139

Einstein's Energy: 140 || Good Ol' Consciousness: 146

Post-Neo-Vedanta: 150

Chapter 8 – Levels of Dark Variables: 155

An Onion Universe: 157 || The Metric of Time: 162 || Gods' Eye: 169

Chapter 9 – The End: 175

Free-Will(ish): 178 || Your Model: 182 || Data Consumption: 187

An Age of Social Science: 191

Chapter 0

Pre-Intro

Beginning. Foreword. Preface. Acknowledgements.

Why not have a Chapter-Zero, instead of the seemingly disorganized and quite confusing pages at a book's beginning? There's no reason why. Personally, I always get lost there. And, how-come such a critical number and concept is dropped when we teach children how to count? Why does it always have to start with "1"? That is because numbers are defined by "0", by nothingness or nonexistence, as the strong opposite of somethingness. Thus, "0" can't be mentioned in the same breath with the 'ordinary' positive-integers. We also don't teach our kids how to count backwards, certainly not at first. That is since the 'arrow-of-time' was (allegedly) shot forward; and also, because we wish our descendants to never-ever encounter anything that is negative. But rest assured, the 'negatives' give meaning to the 'positives' as well, yet in a weaker sense than the opposite-void itself. That is nature's nature. Everything is knotted and active in-between two delineating relatives, a stronger and a weaker one. It may as well be called – *Relativity* (on-steroids).

    A beginning is the time for taking the most delicate care the reader is entertained; especially when considering the current humans' attention-span, not at its best of shapes. But if you are here, it means you are curious, and plausibly, maybe, I hope, care for **Truth**. But even if cared for, perhaps now you ask: "What is truth?" Truth is the-grandest-list of **Facts**, a (dry) physicist would answer. "And facts," you inquire, "what are those exactly?" These are **Events**, or, more precisely even, events that are always of **Interactions**[1]. This is the state of physicists' knowledge and beliefs. Everything fluctuates, oscillates, interacts; and all that seems solid, stable, essentially vibrates. You might come to this conclusion, realization even, once we'll wander in the unintuitively realms of the ('big') Cosmos and the ('small') Quantum. Honestly, above all, I hope you'll enjoy and get intrigued by some of the subjects covered in this book. Intellectual entertainment, a mind and soul's joy, is of great value and should not be dismissed as having no effect on personal growth. Any attention given, anything that gets you "Closer to Truth"[2] – ought to be pursued.

Foreword

"Why should you read this?", is said to be the content of this section, supposedly. It surely wasn't meant to be written by me. But it is, since it is very-hard to find a physicist who'll openly speak of psychedelics. Still, I'll try and make the most of it. The best selling-point I can think of, keeping you involved, is that it seems as if *Psychedelics* are having a moment (it's mid-2021), and they're going (somewhat) mainstream. These are splendid news, indeed, serving two main benefits, I believe. The first is in the acute field of

---

[1] Physicists will say it's only Carlo Rovelli's view (*"The Relational Interpretation"* (30.9.2021)), but it's all of physics.

[2] This is also the name of an educating TV-and-YouTube series, where thinkers debate the vital ideas of existence.

Mental Health. There are substances out there our bodies and brains have receptors for, that some past-humans have arbitrarily decided to completely ban. Not to research, not to educate. Totally prohibit. But this is changing, and that is great. We may finally possess fitter tools for taking better care of all the Post-Traumatic-Stress-Disordered living among us. Here, in Israel, for quite the obvious reasons, there's an abundance of PTSD(ers) to go around. Come visit! The other door through which I can envision psychedelics barging-into the mainstream, is the door of the discipline of *Physics*. Probably, while surfing a 'Wave of Consciousness'.

That is the junction where psychedelics and physics meet – "In our heads" (some might call it). Consciousness is already an integral part of physics, and at the core of most philosophies of existence. Again, two drivers are there. The first is the fact there's an experience of awareness to events which every human feel, and some physicality to it must be somehow physically-evident. Measurable using brain-scans, they believe. The second intersection between consciousness' studies and physics is the definite motivation for this book. For a while now, for the last 100 years at least, consciousness is slowly creeping into the realm of physics' models and theories. Sluggishly, at a snail's velocity, our physicists find themselves in-agreement with some of the most ancient philosophies that humanity has ever conceived. I will intrigue you with the following 'teaser-trailer'. It seems as if the Observer and the Observed are physically-really-united, also in-time; and, that on the most fundamental of levels there are only Potentials (Waves) and Actuals (Data). Consciousness, as it'll be presented, is a phenomenon of both the waves and data.

Preface

I do-not know how persuasive the last two sentences were; but it is said that here this effort's story is supposed to be told-of. My interest in physics arose in the non-common fashion, I can only guess (lacking the full statistics). Consciousness was where I've started my research into reality. The year I consider as 'ground-zero', when it comes to a deeper realization of what's real, is 2008. That year I took a long break in-between my bachelor and master degrees, during which I had visited India and Australia, again. This time, I knew what I was looking for. In Varanasi, India, I've officially begun my journey into consciousness from the perspective of Hinduism. More accurately – *Vedanta*. It can even be, and commonly in the West it is called – *Yoga*. I don't mind, whichever word you like. The journey that started in India had led me to a second tour in Australia. Unlike the first time, which was right after my military service (2003), now my plans were more focused on experiencing a full season of outdoor-raves, and psychedelics. And so-it-was.

That miraculous year has ended with a book, a psychedelic-novel, which I wrote in 2012 and published in 2018. It had received the name – "That's Not Thinking". In Hebrew it sounds much-better, believe me. Consciousness and psychedelics were at its epicenter, and explored in a storytelling's style. It was more fitting for the insights I was attempting to channel, and the experiences I had while studying Eastern-Philosophies and later experimenting with my brain. I like that book more-and-more as time goes by, this novel-like self-growth material. Just prior to me being preoccupied with its writing, a BBC documentary series, "Horizon", aired the episode – "The Hunt for the Higgs". That was where my interest in physics was intrigued. By then I was already a professional in Statistical Learning, and understanding what they were talking about, how the Higgs Boson is searched-for, gave me an incentive to deepen into physics. By the first hour of googling it was clear I was on-a-hunt for no-boson. The physical nature of consciousness, beyond its philosophical or metaphysical logic (which my first book dealt with), became my point of interest. That day's inspiration is manifested in what you read here-and-now.

I did not expect to find consciousness at physics' core; and surely, I did not assume that our leading physicists are more religious and dogmatic than I then-was. But it all had happened, and the last 10 years brought me here. Luckily, the landscape of physics has gained much traction and popularity during the last decade. And during the last couple of years even more so. The months that humanity had just recently spent in (COVID-19) lockdowns, led to a huge spike in the scientific content available, for good-and-bad. This leads to more traction, leading to more content, and so on. It almost feels as if we're amid a social-process of democratizing physics. Some, few, even struggle to save it from itself.

Acknowledgements

First and foremost, the many (white) men and one-and-a-half women mentioned in the following pages, must be acknowledged. Without them, it was up to somebody else to dirty their minds and hands[1]. But they were the ones who did, and we all should pay them respect. Some even gave their lives, so that we all can know what's real. Deservingly they are called – The Founders (which doesn't mean they are free of grave blunders).

Secondly, and I have already cited such one, I wish to assert my gratitude to all those TV and YouTube channels that took it upon themselves to educate us, democratize physics, and teach science in general. The best of them all is – "3Blue1Brown". The "Big Think" portal, and the "Symmetry" and "Quanta" Magazines, are also doing a great job, in my opinion. They keep me fascinated and open-minded. Lastly, I wish to deeply thank those who allow the availability of every

---

[1] Don't assume that without The Founders we would all be living in some dark caves. That is not how life works.

book, old-or-new, for the price of a click-or-two. Without them I couldn't, so-easily, access and research the books of The Founders. Thx!

A clarification I wish to add. These words are not meant to say what's true, nor the whole truth or nothing but the truth (so help us God); and, it certainly does not intend to encourage anyone to experiment with drugs. The purpose of this work is to present you with why and what physicists believe-in; outline how physicists think and how (bad) they are with philosophy and language. As this read progresses, the ideas illustrated will become more 'meta', and beyond the commonly accepted concepts of what is really 'real'. Finally, those beliefs of our top-physicists, and of-mine, with some of the most conclusive experimental results – will all be merged into a simple yet powerful metaphysical vision.

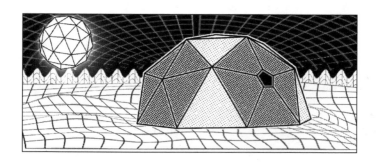

Chapter 1

An Intro

What. Why. Who. How. And some definitions.

Like they do in Hollywood, I'm going to start at the final shot, with my concluding thought. It seems as if the discipline of Physics got blundered with itself at a crucial crossroads, and now what's required is some-kind-of 'New Physics', prophesied to be delivered by some-kind-of 'New Einstein'. These are not my own words, nor a wishful thought. This conversation is mainstream among leading physicists. Moreover, I have come to believe that this scientific evolution, not a revolution, will inevitably require an assimilation of some non-purely-physical phenomena, mainly that of *Consciousness*, into both the new theories and their *Metaphysics*. Dead and gone, this I certainly know, are the days of dismissing such a 'fundamental force' of nature with null statements in the spirit of – "It is just your brain playing tricks on you." Humans' self-witnessing and interaction with nature is no longer principally overlooked, and scientific reasoning is demanded. So, what's so deeply erroneous with modern physics? How can it be so widely accepted as mostly-fruitful-yet-very-stuck? And, how is it connected to our concepts of **Meaning**[1] and *Time*? Also, what are the developments the coming days may bring; what paths can some of the newest theories, the oldest Eastern philosophies, and even psychedelic experiences, point science next-at? That is what the following pages will mainly contain.

---

1 Seems like we humans, and only us, have some obsession with Meaning. Maybe some dogs also.

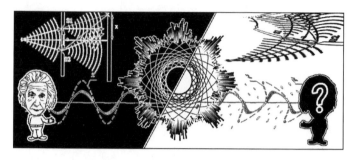

Important to note, that the bit-haughty work here is not anti-science, and not against physicists or physics (which is just another word for physicists). I adore them, and even considered a scientist myself. I am pro-science, pro-facts; and even pro-truth, dare I say. Bearing that in mind, I advertise this work as scientific and logical, and an analysis which also offers a clear view into the nature of theories and models themselves and on overall processes in science. History, the grandest list of facts, is always there to provide us with some (in-hindsight) perspectives. Therefore, physics' stories will be scattered all over the place. Everything I have compiled here to make my case, can also be easily seen as a review of the *History of Physics*. "But wait, who the-hell are you?", I know I would ask.

As a researcher, I belong to both the rapidly declining discipline of *Economics* and the booming professional network of Data Scientists and Machine Learning Engineers. In the 1990s, people such as myself used to be called – **Quants**. It had nothing to do with the *Quantum*. We were defined as such since our job was to quantitatively analyze the capital markets, the *Big Data* of the then. Nowadays, most of us are so-heavily focused on the Internet. I was 'trained to think' under the widespread umbrella of economic analysis, where they taught me to investigate questions and new hypotheses, i.e., theorize *Models* to be statistically tested against real-world **Data**. My thesis offered an analysis of Central Banks in an economic environment of Bubbles & Crashes. As for what I do with it to earn a living, I am of those to be blamed for all the annoying ads we encounter online, in apps and websites. There's no glory there, but some, sometimes, even call the purely statistical models I build (to maximize revenues from ads) – 'AI' (*Artificial Intelligence*). Later it will be demonstrated how **Predictive Models**, and their true and only nature, are decisive in my analysis. In a sentence: Few models are considered by many physicists as 'real things' and not the mere approximation-functions such tools truly are. As a predictive modeling architect, I have some pretty solid thoughts on the matter.

Similarly, there is another group of (somewhat wild) researchers one can clearly identify me with. **Psychonauts** (psychē = 'soul/spirit/mind' & naútē = 'sailor/navigator'). Psychonauts can be loosely defined as those who investigate consciousness ("which is?", readers I aim for will ask). Such navigation within and exploration of consciousness is done by some form of meditating, with or without consuming psychedelic substances but while always keeping in mind the logics derived by consciousness' researchers of the past. It is tempting to have ideas about the nature of consciousness, or even question its existence. But, please, don't go inventing the wheel. Meditation and its virtues were and still are taught by sages and gurus, such as Shankara or Ramakrishna or Patañjali, and it is commonly viewed as a tool for looking inwards, observing **The Self**. *Vedic Philosophy* is the oldest, and certainly the most well-established, research discipline focusing on the investigation of our-true-selves. In the West it is more frequently known as – Yoga. In sadness I say, that I find western-yoga, very generally speaking, much more misleading than insightful. My quick advice is to read yoga, mainly **Jñāna Yoga**, prior to exercising it. By the way, the exact same advice goes for Psychedelics, and drugs in general.

Utilizing Psychedelics, by consuming Psilocybin or 5-MeO-DMT or Mescaline, just as examples, alters the experience of existence while under the influence and has long lasting effects on the view of reality, including its physics. Psychedelics are all-the-rage in the field of Mental Health; and, as time goes by, more physicists are coming out of the 'psychedelic closet'. Just as *Cannabis* has already did, psychedelics are going mainstream, and new psychonauts are born daily. Psychonauts 'go-on-trips' for the sake of exploring the inner-workings of the *Mind* (another term soon to be defined), and not just-for-fun. Just-for-fun consumers of (what's also called) 'Hallucinogens', attempting to hallucinate some cool visuals while at a trance party (pardon my cliché), are 'missing' psychedelics' purpose in my opinion. That's since 'Psychonautism' is scientific in nature. It's just wrong to label psychedelic experiences as – hallucinating (or intoxicated) brains. I find the very-opposite seems true. If anything, psychedelic experiences are physics' reveling, since during those moments our minds are processing irregular amounts and shapes of data. Could very well be, that psychedelics expose our interaction with, within, reality.

The viewpoint of the **Veda(s)** and of **Vedanta** ("Vedas' conclusions", from Sanskrit), on what consciousness is, will serve as the presumption of this research, and I think it's for a good reason. Vedanta was originated in India, millennia before it was the nation I have visited again and again. It has not only given us the physical and philosophical practice of Yoga ("Unity", from Sanskrit), it has also introduced the concept of *Zero*, of **Nothingness**, to human thinking and mathematics. More than a decade I've invested learning the fascinating development of Hindu Philosophy. Even though I experimented with psychedelics years before, no mushrooms or toads were harmed during that part of the journey. Firstly, I understood the powerful logical reasoning and claims on the reality of The-Self made by *Advaita Vedanta*. And, eventually, I've become a **Neo-Vedantist** in the spirit of the lectures and writings of Swami Vivekananda. In a sense, psychedelics had served as an aid (yet not a necessity) to the practice and understanding of Vedanta and yoga. Psychedelics are magnifying consciousness, surfacing and helping in realizing it, which is the ultimate destination of researchers of The-Self. I admit, I was convinced, but if we agree, and we should, that both *Pantheism* and *Panpsychism* are some next-of-kin to Vedanta, it seems as if many of the most influential scientists have also been.

When surveying for the philosophical and religious views of our leading physicists, Pantheism and Panpsychism are clearly popular, and some more. They both preach for consciousness' fundamentality; and, like psychedelics, these metaphysical-philosophies are also having a moment. That is a very fresh development, dates back only to the previous decade. Its main driver being the fact that consciousness has gained a major role in the interpretations of the most successful theories and models of physics; and, some experiments are hard to explain without some role to it. I'll show you exactly which, and why I find it almost inevitable that 'psychedelic-ideas', factual feelings and observations from such experiences, will eventually find their way into physics' labs.

In this work, one will find the story of what we're told physicists think and know of the fundamentals of existence. This task is no-easy; with the grip of *Populism*, during the *Post-Truth* era, tightening around our necks; and, the growing amounts of the false-facts (*Fake-News*) that're daily bombarded on us. And that's not all. These words are written during the second year of the SARS-CoV-2 'pandemic', where the field of medicine, and science in-general, have suffered a major blow to their credibility, being so-very evidently politicized. Conspiracy theories are a business drawing plenty of (online) traffic, which I find upsetting. To confront this obstacle head-on, and to avoid expected pitfalls, some of the most important theories, experiments, scientific debates, several controversies of physics and even a conspiracy (there is at-least one), will be explored and interpreted and reinterpreted. This will be done while reflecting on some of the cutting-edge ideas in physics, alongside the most established notions of consciousness. I also take to heart the conclusion of my own work. I read the fact that I'm a multidisciplinary researcher, who found great interest in metaphysics and the **Psychedelic Experience**, while probing for reasonings in mainstream physics; as a 'hint' that I do have something to add, and should write these words. I wish to share with other seekers of reality, spiritualists and scientists alike, my two-fold journey into what is really-real, both the-physical and the-not.

Although this work is more than fairly scientific, these words are primarily meant for non-physicists and for physicists-to-be. As its title indicates, it was written for psychonauts, for those attempting to sail and navigate within the vast 'sea of mind and consciousness'; whom I think must have a greater impact on physics, and science as-a-whole. My hope is that some of them will gain a better understanding of the physical reality, before they enter the psychedelic domain and start experimenting with their brains. I wish to clarify some of the experiences they'll undergo, and give them a deeper, real, physical and mental meaning, far-beyond the mere simplification of 'hallucinations'. I hope that such educated-psychedelic-experiences will grow their interest in physics, and even in truth. Psychedelic experiences should be viewed as (mini) self-experiments, I think, that lead to (maxi) self-evolution, from my personal experience. Again, these are not party-drugs.

This product should also serve as an easy-to-follow read for whoever is in search of a multidisciplinary analysis on the state of physics, and the junction where our physicists had 'hit-the-wall' of consciousness (and got utterly blundered). As mentioned, the history of the evolution of physics is of significance, and those who wish to get a very-condensed version of it, will find it here. Finally, I believe this book should draw 'Yogis', spiritual-seekers, and people familiar with the philosophies of *Buddhism* and Vedanta, to study, and maybe influence, the field of physics. You have much to add. Complementary, established scientists who find it hard to digest the growing focus on consciousness in physics, the experimental evidence and several fresh theories including it, will also find some interest here (I'm pretty sure). Now, let's start talking some physics.

State of Things

Mainstream physics is in a state of 'blunder', a term many physicists, quite surprisingly, usually associate (alongside "genius") with Einstein. I'll elaborate on that part of the story in the next chapter. It's actually quite long, the tales of Einstein's blunders. Still, it doesn't mean he is not 'The JOAT' (The-Jew-Of-All-Time). Don't trust me that physicists are, as a group, blundered. Read, "We Have No Idea" by Jorge Cham & Daniel Whiteson (2017). It is good stuff; I highly recommend it. Many of their insights you will find here, but they just do it much better. This state of blunder spans from the scientific use and validity, of notions such as *Space*, Time, and *Matter*. All the way down (and up), to what we are being told are the presumed 'Beables'[1]. I'm sure you already know that – "there is no spoon". Everything is *Particles* structured into *Atoms* and then *Molecules*, and up-up it goes.

Please don't be fooled by the shrinkage of machines or the universe's finer images they've computed. Don't even be impressed by the fact, that only this year it was (again) announced, after more than some 100 years of no direct proof, that the most famous of theories has finally been proven (subject-to "heroic computation"). By the end here you'll understand exactly what I meant in that last sentence; but, for now, let us first see why *Fundamental Physics* is so blundered, and why their blunders present an opportunity.

For physicists, existence is dichotomy divided into two. This is also where one of their most-acute blunders lurks. Physics' models separately analyze the 'Big' (Classic Objects, like you and your phone, or the Moon) from the 'Small' (*Quantum Particles*, such as the *Electron*). Where does 'small' end and 'big' begin, and by which mechanism precisely? That's a great question; especially when considering that measuring

---

1   That is John Bell's proposition as a replacement to the problematic term 'Observables', in quantum physics.

devices are 'big', not quantum, yet still are made of *Quanta* (quantum's plural). In addition, and many readers might not be aware, but physicists assuredly-assert that more than 95% of all existence, of the Universe, is 'Dark'. That is their way of saying: "It does not interact with *Light*, thus it's unobservable to our (big) measuring devices." They just have no idea what it is, but they're still sure it is real. This is a big-blunder, one of cosmic scale. Furthermore (and I find their confidence as mind-blowing), physicists also maintain these above-95%, of 'things' they cannot see and know almost-nothing-of, are comprised of *Dark Matter* (27%) and *Dark Energy* (68%). And finally, about both these two 'darks', physicists advocate (with zero hesitation): "We can't see it, but we know it is there". Sounds a bit like what a Rabbi would say, doesn't it? Deducing what exists from indirect evidences is a perfectly-sound scientific methodology, but I'm confident it should be applicable with extreme care. We will see so-many, endless occasions, where scientists are so-sure of something, confident of some 'facts', while still lacking any direct observations to back them up.

Here I must quote Nicola Tesla, the dude with the car's name, who (apparently) said: "The day science begins to study **Non-Physical Phenomena**, it will make more progress in one decade than in all the previous centuries of its existence". I know Nicola was aware of the thousands of years that metaphysics received attention in the East, since in 1893 he debated with Vivekananda, the bearer of **The Movement for Self-Knowledge** to America. Also, so it seems (for reasons I'll present), that the decade he pointed to is the current. Now here, just as a balancing act, I'll add an opposite and a representing quote (1977) by Steven Weinberg, who passed away just this last July: "Our mistake is not that we take our theories too seriously, but we do not take them seriously enough. It is always hard to realize that these numbers and equations we play with at our desks have something to do with the real world." See? Weinberg, who was extremely influential, thought they do not take themselves seriously enough. Personally, I think that most (if not all) humans should-not take themselves too-seriously. I even believe it can be proven physically.

Now a question I ask. Why should you, a busy person in a competitive world, spend your precious time reading-physics (from an economist no less!), instead of scrolling over photos and 'stories' or swiping between potential matches? My answer is, that knowing what's real makes a person stronger and fitter for post-truth digital-societies. Stronger, in the sense of becoming less inclined to being over-influenced, or even manipulated, by others who simply just know better; and, more balanced in reactions, since having proper proportions of what's (really) real. In the final chapter it'll be demonstrated, where the many ideas in this wide analysis will lead to a practicality for everyone's every-day. The subjects of *Free-Will*, and that of Time, will be tackled with the insights drawn here and will probably be seen in a fresh light. It is a cornerstone of my conclusion regarding the physical validity of self-non-seriousness; the fact all the data points at a very limited existence of free-will (still less limited than that of animals), and of time as well. And finally, **Attention**, our own human attention, will also receive some attention here, and shall assume a much more profound and deeper role than is commonly been-given.

This That You Read

Some of the words here are *Italicized*, while very few are **Bold**. Bold words will emphasize that I, and probably a few others, find them crucial, and a subject or idea worth contemplating upon. Italicized words will indicate that the subject or term is, objectively, of great interest and meaning to the scientific and philosophical community. I advise the reader to google them, if one wishes to go down the 'Internet's Rabbit Hole' of videos and blogs and papers on the matter. But, and this you should already know, every word or term, even half-a-string or just a single character, is **googleable**. Since this read is meant to be as user friendly and accessible as possible to the laywomen or men, it will include no graphs and just a single equation (the most popular one you are already familiar with). I know scientists couldn't care less for words, since they speak in the (very limiting) language of math, and math alone. My hope is to persuade you this is yet another reason for their ongoing and deepening blunders. In that sense, I almost feel like – physics is the new economics.

The very general structure is as follows. At ground zero I would lay down some basic foundations of definitions. This will be quick and painless, I hope. If it does cause you pain, trust me, it is not because of something I am doing. All these terms will be clarified and even proven over the course of this book. These are there to always there remain, for them to maybe be-visited again. Then, it will start with an overview of the history of the human understanding of physics and (the philosophy of) *Nature*. During and after that, a couple of not-so-minor subjects, such as Space, Time, Matter, and *Energy*, will have to be rethought. It is not a rethinking done by myself; these are evidently being rethought by leading scientists and physicists. It will be established by summarizing and freshly interpreting some of the most impactful experimental results, from the last 220 years of advancements in both physics and its philosophy. Once the history will be surveyed, it will become possible to build old and new ideas on-top-of the facts that most agree upon. As this work progresses, the insights and analogies will become more stimulating. It all will be sprinkled with ideas, that jointly may-seem as yet-another – Theory of Everything.

As mentioned, Vedantic Philosophy will be built-upon and fully integrated into the physics, in the spirit of works of researchers such as Deepak Chopra & Menas Kafatos. "Physics does not like them," I've been told, yet I couldn't care less. Them both must be mentioned here, even though I have my own lucid-visions and educated-perceptions of (what we Vedantists call) – **Manifestation**. Another piece of research that I think might support my work, and I'll tell much of, is – "Space Is All There Is" by Shlomo Barak (2020). I've found his work while searching for papers that seem to conclude with the same main insight as mine, that Space (and Consciousness) is all that's required. I was in awe by Shlomo's work of 25 years, and won't be surprised if he'll receive a Nobel for it. It seems he has unified the physics of the 'big' and 'small', by sharpening Einstein's work.

Now I'll risk of sounding a-bit arrogant, but, although Dr. Barak and little-minor-me certainly do not agree on everything, it seems that his ideas, his *Geometry of the Universe*, may serve as the mathematics of my (indirect-and-meta) physical analysis. Still, there is one aspect where it seems I truly read and interpret the facts simply just the opposite than everyone else, including Barak and Einstein. I'll show you where and what and how, and you'll be the judge of it. Overall, I'll try to lay my thoughts in a fashion that even the reader who didn't absorb all the bits & bytes of what physicists preach for, explicitly or not, will come to grasp with where contemporary physics stands on the hazy matters of elementary particles and spacetime. As advertised, psychedelics and Vedanta will creep into the physics and enlighten it. Hopefully, it'll allow for more established disciplines of consciousness' studies to feed the science with new and disruptive notions. So please join me on this journey of mine, a journey from the 'Quantum Realm' to Consciousness and back, where I'll strongly try to advocate – **Why Science and Psychedelics Go Hand-In-Hand**.

Some Definitions

We all hear physics' words and scientific jargon in everyday life. Popular-pseudo-science is a booming business; its content is widely spreading and being used to sell lots of stuff. That is why we are here, let's be real. In addition, the New-Age and spiritual literature is filled with terms that Eastern Philosophies, like Buddhism and Hinduism, have been dealing with for centuries. However, it seems as if words have lost all meaning, and only buzzwords remain. I wish to try and right this wrong from the-get-go, and simplify some terms to be explained and used throughout this book. I believe even if it will be the only section readers consume here, without the later storytelling and elaborations and the deeper-dives, they will be quite benefited. I'm also of those believing that words have a very-precise meaning of what they describe, hence should be used with great care. A word of warning to readers who, for them, this is the 'first walk in the park' or 'first time at the beach' (of physics). Please, don't be alarmed if all these terms sound like gibberish, now. They may seem cracked or unclear, but that's the purpose of the effort here. Making you comfortable with the following terms, so that you're never intimidated by the words of 'an expert' (physicist or economist) – is my highest of goals. Now, let's commence.

General Relativity – Einstein's 1915 reimagination of Space and Time, into *Spacetime*, i.e., the 'arena' where all matter exists and moves within. With this one he reshaped our understanding of *Gravity* and the Cosmos. This is still the dominating theory of the large objects, such as the Earth or Sun or Blackholes, and how such massive planets and stars interact with, shape, and move within, Spacetime. **Spacetime**

Matter will be at the heart of this work, as it was for Shlomo Barak. General Relativity was the 'second-child' of Einstein's Relativity, but all agree (and he thought so) that this theory is his favorite-one.

Special Relativity – Conceived 10 years prior to the General, where Einstein coined the most popular mathematical expression of all – $E=MC^2$. He did so while analyzing the changes in *Mass* and Energy of (non-accelerating) moving objects. Here he completely reshaped our understanding of the *Flow of Time*. Time's nature, and its 'creation', will also be examined here. The relationship among space, light, and time, are of existence's essence. It's quite unbelievable, but it is hard to find when exactly $E=MC^2$ was proven.

Quantum Mechanics – The dominating theory of the behavior of the 'small' pieces, the Quanta. Everything including everything, including atoms themselves, are made of tiny, the smallest we know of, 'point-like' (and it's important) *Elementary Fundamental Particles*. The 'packet' of light, a 'piece' of the (electromagnetic) wave (or field) of light, for example, is a point-particle called – *Photon*. It was also Einstein who taught us it is real, and not just somebody else's mathematics. Yup, Albert sure made some impact on the discipline.

The Standard Model – The result of the *Quantum Revolution*, the quest of physicists to quantize everything in nature; plus, the work of Richard Phillips Feynman, formalizer of *Quantum Electrodynamics*. This gave rise to *Quantum Field Theory*, which is the Standard Model (SM). Currently, in the SM, there are two sets of three couples of matter-particles (Fermions), and five more carriers-of-forces-particles (Bosons), with the sixth being yet another major blunder. The need to quantize Gravity, to observe a piece of gravity's force, might just break the whole damn thing up. Ah, and there is also a counter *Antiparticle* to almost every one of these 'Non-Antis'. Relativity-on-steroids. While I'm writing this, a new and unexpected force, that can shine a new light on it all, may have been discovered[1].

The Multiverse – Just one example of the many *Metaphysics of Quantum Mechanics*. Quantum Field Theory, and most of its experiments, point to some extremely unintuitive and (so-called) 'weird' behavior of matter and nature on the subatomic-levels. Therefore, providing interpretations, attempting to give a logical meaning to the observations, is a fast-growing business. The Multiverse is yet another one of those, quite a popular one, and I think it's the best example of how physicists view their mathematics as a real thing.

Dark Matter – Originally known as 'Missing Mass', Dark Matter's existence was first inferred in 1933, when it was discovered that the mass of all the stars in the Coma cluster of galaxies provided only (about) 1% of the mass needed to keep those galaxies intact. Dark Matter is suspected to be the glue that binds celestial objects together. By the 1970s it had received the label of 'real', and has been confusing physicists ever since. 'Dark', basically means – can't be (directly) seen. They observe something unexplainable, and just add 'dark' to it. For now, physicists still have no model nor idea what's

---

1   The name of the paper where they've published it – "Test of lepton universality in beauty-quark decays" (2021).

happening there; and, the entire concept of Dark Matter puts some big question-marks on Einstein's General Relativity. It will be shown how 'Dark' can be replaced with 'Hidden', and vice-versa.

Dark Energy – Same case more-or-less, only this one is not for an unseen special matter that holds galaxies and clusters of them together. This other-dark is used for the invisible, most-mysterious and all-pervading, strongest force in nature. This energetic-repulsive force is (said to be) trusted with the task of making sure that the universe expends, being pushed around, all the time and in a non-constant velocity. As an idea, Dark Energy is much younger (and abstract) than its matter-dark-brother. It was suggested only in 1998, to explain the deformation (*Redshift*), caused by the travel-time in an expanding universe, of light that's coming from different locations in the cosmos. These efforts are generally very-much limited by the deep-difficulty to accurately measure astronomical distances.

String Theory – The attempt to unify General Relativity with Quantum Mechanics into a single Theory of Everything (ToE); to merge the 'big' and the 'small' into a single model. In this theory, gravity emerges from a particular vibrating string. Gravity is built-in. It's some nice math, but currently it seems as a colossal failure. Let's wish them all good-luck.

Wave – The thing in water that surfers know how to enjoy. Also, as Thomas Yung discovered in 1801, it is the nature of light. "What?!", the attentive reader asks, "you've just explained that light is a particle named a Photon!" Well, yes, Wave-Particle Duality is at the core of physics and the root of the field's deepest controversies. Maybe, it is their biggest philosophical blunder, the one I think got over-hyped and totally bewildered. Speaking of waves, these are quite evident in some psychedelic experiences. Everything becomes wavy for me; all I see (including when I close my eyes) once I'm consuming and consumed-by **Psilocybin**, seems flowing and filled with waves. From the first experience to the last, more than 15 years apart, all matter seems to lose its appeared steadiness and is suddenly 'alive'. One can say I was hallucinating; I say I got a better glimpse of reality.

Wave Function – The math/statistical tool underlying Quantum Mechanics, and the number-one root-cause for the Wave-Particle Duality's philosophical blunder. Louis de Broglie, in his 1924 PhD thesis, suggested that all matter, just like the Photon, has wave properties. This a powerful tool for deriving probabilities for measurements' outcomes. In a Wave Function, where the wave is high, there's a higher probability to find a particle.

Mind – A computerlike organ, certainly the processor of incoming and internal data, we call – the Brain. Can also be thought of as the arena of human, and probably several other creatures, experiences. This should never-ever be confused with – Consciousness.

Consciousness – The Awareness, the (Ultimate) *Observer* and Witness, of experiences and events and even of the mind-work itself. It is much closer to the common spiritual, and even religious, notions of – **Soul**. 'The Observer' is an extremely loaded subject for physicists and Vedantists alike, one that I shall tackle from all angles and vantage points. Some contemporary (and Western) theories of consciousness, in the likes of *Integrated Information Theory*, attempt to quantify the phenomena and allow it concrete place in labs. This is an important development. 'Levels of Consciousness' (or of **Synchronization**) and 'Levels of Entanglement', are just examples for the directions I would suggest following.

Entanglement – The term used to describe the state of association between two or more discrete physical entities once-interacted, and even though they might be at very extreme distances from each other, post-interaction. Quantum Mechanics' math predicted, that any following interaction with one of the entangled particles will instantly (even faster than the speed of light – the C in $E=MC^2$) influence the other particle(s). *Bell's Theorem* of 1964 showed physicists how to test entanglement, and experiments proved it over and over since the 1970-80s. It clearly exhibits that the universe communicates faster than the speed of light. Einstein (& Co.) had found it, and got full on blundered over this one. He took it personally, since entanglement strongly challenges his two Theories of Relativity.

Blackhole – Predicted to exist by General Relativity's math, coined by John Wheeler in 1967, and firstly seen in 2019. 'Seen' is not the right word here, since a blackhole is, well, black, trapping most light passing nearby. Physicists 'spot' a blackhole by analyzing the activity of space and matter and light around areas where a blackhole is suspected to reside. This (very limited) empirical reasoning has led our physicists, more than once, to wrong conclusions (such as the existence of Dark Matter). Blackholes are now suspected to be at the center of most galaxies, as there is one most-probably at the heart of our *Milky Way*.

Information – This describes what can happen, the entire 'space' (as mathematicians and physicists say) of all possible measurements' results. When we roll a die, for example, the information available is that the result will be, and only be, a whole positive number between, and including, 1 through 6. The Wave Function is the physics' math-tool that holds this information. Physicists repeatedly confuse between Information and Data.

Data – The outcome of a *Measurement*, or any other event and interaction, that results with the creation of a certain objective fact. Following the previous example, the number showing on-top of a just rolled die, is just a piece of data everyone can agree upon. You won't believe what physicists are calling the die while it is still rolling, and its implication. *Causality*, causation, cause and effect, are data-points, above anything else. Since Special Relativity, it is known that different observers won't agree on the order of events, yet Causality is a fact that all can agree on. Again, they truly confuse Data with Information.

Well, I hope that wasn't too harsh, all those terms above. Some were probably new, while others were not what you thought they were. Don't you worry, all will be deeply-explored and brightly-clarified. So, now that we are through with the Intro, and there's a place you can go back to if a term or concept is lost in-elaborations, let us start with the real-deal – the *History of Physics*. Only history can explain how we've reached the point where physicists need to remind themselves and their colleges, that molecules, atoms and quanta, were just metaphysics, nothing more than a vision or imagination, up until the second they were deemed-real. Also, history will clarify how physicists got so blundered.

Chapter 2

Some History

*Many accidents. Some blunders.
Mostly white males.*

I'm glad you are still here, and I didn't drive you away. But, after I probably did (already) place us in some-sort-of disagreement, and for the sake of exhibiting a-bit-of humility, I'll begin with a recommendation for those truly interested in the history of physics. Read "Reality Is Not What It Seems: The Journey to Quantum Gravity" by Carlo Rovelli (2016). With him you'll become familiar with the evolution of our understanding of what there is, or why physicists have-faith in the things they believe-in; and, you will also witness the fixation, the obsession even, of our most popular physicists for *Gravity Quantization*. Rovelli is a great teller-of-physics, and he's also one of those physicists who are out of the 'psychedelic-closet', drawing a direct line between his trips at age 16 and his journey into *Loop Quantum Gravity*. An Italian, what did you expect? They know how-to-live best.

Unlike Carlo, I wish to tell the story of physics when divided into four discrete eras, and while going backwards. I also permit myself to be much-more critical, since it is not my (professional) well I'll be 'spitting-in'. No worries, no bodily fluids will be scattered from within these pages; it's just a phrase. Still, this work is being compiled in mid-2021, during the (bizarre) days of *COVID-19*. We are going to start with the current millennium, with the unexpected discoveries and the desired confirmations of the last 20 years. Both Quantum Mechanics and General Relativity saw the bearing fruits of billions of dollars spent on huge and long laboratories, meant to further confirm themselves. Of no less importance, is the fact that during those 20 years we have witnessed a tectonic shift in physics' attitude towards consciousness, this work's root-cause. The second era I wish to 'quantize' as a standalone piece, started with the 1949 coining of the term "The Big Bang", and ended with the 1998 (somewhat blundering) discovery of the accelerated expansion of the cosmos (due to Dark Energy), since that Big Bang. "But, what matter is expanding? And expanding just where exactly?" These are some fine questions you ask there!

The third era I will package together, started with Max Planck's mathematical hack to solve the *Ultraviolet Catastrophe*, that de-facto initiated the Quantum Revolution in 1900; and, concluded with Feynman's formalism of *Quantum Electrodynamics* in 1948, allowing the establishment of the *Standard Model of Particle Physics*. And finally, I'll review the important men and (one) woman, who led to the discoveries made between Sir Isaac Newton's *Laws of Motion and Gravitation*, conceived in 1687; until the 1897 discovery of the Electron, by Sir J.J. Thomson. Just an instant (psychedelic) fun-fact: Arthur Heffter's isolation of Mescaline from the Peyote cactus, the first isolation of a natural psychedelic substance in its pure form, also happened in the year 1897. Now, let the storytelling begin.

From 2000

The last twenty years have been quite interesting for physics, still not so great for (too) many physicists. Nowadays, the growing amount of times-per-physics-zoom-conference that someone says: "A new Einstein might be needed here," I think, simply says it all. It seems that almost every new discovery in the field of *Particle Physics* can jeopardize the validity, or, at-minimum the accuracy, of the entire Standard Model (SM). Nevertheless, two major (and very expensive) experiments, which also exhibit how General Relativity and Quantum Mechanics so-differ in nature, were finally concluded and deemed successful and useful, so it's claimed. One thing can certainly be said, physicists' scientific announcements are no longer taken at face-value, which is a direct result of the following.

Another important progression that physics has undergone, was the democratization of the field. The Internet, practically lifting the gates and breaking the walls of academia's monopoly, is allowing all an access to *Knowledge* that previously was available, mostly, to the privileged. It is not only that YouTube and Twitter are filled with hours upon hours of content regarding physics, which I find as very impactful. I know for a fact, and I'm a living example, that the online availability of the most momentous books and papers in physics, past and present, is a game changer. These works can now be directly consumed, with ease, and be openly criticized without the mediation of rock-star-physicists (who're presently producing more content than an Instagram 'celebrity'). I pity the 'Physicelebs'; they're in quite a pickle. Knowledge that was long held by a few, is now common, if one is only interested in it. Know that you should only be attentive, should only give it some attention and allow it time to sink-in. I advise you to hesitate not; because, as it is prayed by the greatest women in Frank Herbert's **Duniverse** (1965) – "Fear is the Mind-Killer."

Let's start with the *Higgs Boson*, which you should already know is a force particle, since it's a boson (and they didn't call it – the Higgs Fermion). Its buzzword-nickname is 'The God Particle'; and if I were a *Quark* or a *Gluon* I would surely be insulted, since it would suggest I'm less godly, which I would not accept. The Higgs was theorized in 1964 by, among others, Peter Higgs, and emerged in 2012 from data collected at the Large Hadron Collider (LHC). I think that "emerged from data" is such a phrasing that should be deeply considered. I'll do so, and will focus on it in some of the following chapters. The LHC is a huge machine, a 27km-circular-tunnel, built underneath the French-Swiss border. I can write an entire book of how economically and socially immoral this project is, but that's a book nobody would read. I can guess, that you guess, that they've already started making promises of what will "emerge from data", that some ten-times-longer (and more than ten-times-more-expensive) tunnel will produce.

A great place to start digging deeper into particles is the Higgs, and not just because it was the last to 'be seen' ("emerge from data"). The Higgs is a representative example of how physicists think about, and (ab)use, math and statistics. By the 1960s, physicists already had some aspiring equations. Anybody can come up with an equation that makes sense. We (imagine I put my economist hat on) do it all the time, writing economic sentences in the language of math. "An average household will spend 40% of its income, and an additional 1000$, on monthly consumption"; or "a banana field corps forty times its trees' number, times labor hours spent, in tons". These're just examples. But, what physicists had was much more than what economists will ever dream of producing. They wrote equations that describe all the interactions between all force and matter particles they theorized to exist. Yet, there was a problem, seemingly. To keep the equations simple and *Symmetric*[1], particles had to have no mass, just like the photon is supposed to have

---

1   A physicists' obsession that Sabine the YouTuber had criticized in her unexciting book – "Lost in Math" (2020).

none. So, for mathematical reasons, above any else, another particle and a new force-mechanics were invented. "But, wait, what is mass?", I know I have tried to ask.

Firstly, mass is not weight. It's connected, there's an equation, but it's not. Weight is just not fundamental, since it changes by location in the universe, and even the building's floor you are on. That is a very important statement – that only what is none-relative truly exists – (both) for the Physicist (and the Vedantist). Mass is a measurement of how much 'fundamental stuff', energy including, something has. But, for an elementary particle, the building-block of matter, what stuff is 'in-it'? Mostly energy, probably. So, mass (as I comprehend what physicists cautiously say in vague words and complicated equations) is *Inertia*. Movement and mass are intertwined, and even for the 'first born' particle, the electron, to have mass, it needs to be 'pushed-around' or (as they call it) 'get-excited'. This is a very deep glimpse into physics, the fact that inertia is at its core. And, if we remember that $E=MC^2$, and accept the clear and widely putative relation between mass and energy, we're allowed to conclude the following. Mass, Energy, Inertia, and Light's Speed, are interlinked, somehow. A sentence about energy. "It is important to realize that in physics today we have no knowledge of what energy is", said the Nobelist Feynman in his 1960 lectures (which are great and found online). In the 60 years since, not much has changed.

The Higgs, unlike the Electron, is a (boson) force-particle, a Quantum of a *Field* that's everywhere, interacting with (almost) everything. The Higgs itself also has mass, just like (almost) all other particles, besides the Photon, are said to have. They say it's interacting with itself. How convenient, isn't it? Force Particles are what Matter Particles 'exchange' when interacted; and, unlike the fermions, bosons can overlap and be in the same place at once. This is why your hand doesn't just go-through the table or phone, because two fermions can't occupy the same space. Very generally, 'physics says' that particles vibrate and oscillate, therefore they are displaying – mass. Notable physicists even claim that this oscillation is what 'creates' time, that it is – time. That is almost the dominating *Paradigm*[1].

How did they 'find in data' the Higgs, and how do particles get 'found', generally? It is going to sound over simplistic, but, physicists are excellent at smashing things up. Just like a child, testing toys' limits by trying to smash and break them apart, same goes for our physicists. The LHC is all about colliding subatomic matter in search of surfacing their constituents, their *Elementary Particles*. As a technicality, it sounds a-bit like the utilization of psychedelics. And, just like the kids with their toys, trying to break stuff into smaller and smaller parts, more energy is required. What is needed is stronger inertia of subatomic-matters' collisions. To find a particle

---

1 A distinct set of concepts or thought patterns, including theories, and research methods.

means to verify it is stably seen (statistically) enough times within trillions of data-points created by trillions of collisions. Again, to spot a particle, it must be seen many times with the same properties, thus deeming it – stable. Here I must mention the *Neutrino* (the partner-particle of the Electron, that will receive its own sub-section later). The Higgs does not interact with it. It doesn't really interact with almost anything, so-they-say, which means we have no idea where its mass results from. Ah, and there are also *Exotic Particles*, very unstable and seen rarely in nature. One such of these, the most stable exotic ever, was recently observed, in July of 2021. They call it *Tcc+_Tetraquark*, and it's said to open the door to all new kinds of matter. Certainly, we live in some very exciting times, exotic even, (at least) for some of us.

The Quantum Theory and its elementary particles will receive much more attention during this read, but now let's please focus on the second major announcement made by our physicists during the last 20 years. That is the 2015 detection of *Gravitational Waves* (GW). This became possible when the LIGO sensed gigantic ripples in Spacetime, said to be caused by two colliding blackholes 1.3 billion light-years away. Yes, physicists engage in storytelling sometimes as well. The LIGO detector consists of two vacuum tube arms, 4km-long each, arranged perpendicularly in an L shape, with two very sensitive lasers meeting at the L's corner. The LIGO can detect GW coming from anywhere, even below, by recording the vibrations in the lasers' extremely tiny meeting point. I suggest a metaphor here, one that I'm about to use much. Imagine the LIGO being a very small lab on a very big rock somewhere on the very bottom of the deepest ocean floor, that's able to detect a brawl between two sharks that erupted on sea level by sensing vibrations traveling in the water itself. This is the LIGO, and these are the sharks' waves.

Unlike the evidence needed to detect anything fundamental on the quantum scale (an abundance of subatomic collisions), tests for General Relativity (GR) and/or Gravity, usually require just a lone evidence, no more than a single observation. This is another example of how these models so acutely differ. GR can be somewhat, partially, confirmed with a one-time experiment, as it was firstly proven in 1919 with a single photo of an eclipse; while QM and Quantum Field Theory (QFT) will always require trillions of observations (data points), in-order-to judge a quantum model as (statistically) – 'real'. The next sentence is of great importance and meaning, so pay attention please. A QM's model shouldn't be 100% correct, just to be **On-Average** (statistically) precise. Nobody knows much about just a single particle; unlike the movement of the Moon, for example, that is easily predicted. QM and QFT, as probabilistic models, can only predict outcomes of many-events, of many-interactions, but never for just a single measurement. That is very similar to the predictive models I build, that can never accurately predict if a person will click an ad or install a browser-add-on, just the **Probability** that many of such a user will behave as my clients prefer them to. It is a crucial point to analyze. On the other hand, GR, or any theory of the 'big' scale, must always be right, and everywhere in the universe. This is also quite a contemporary blunder, since cosmologists are starting to gain images of things in the cosmos that their 'theories scream': "Hi! This shouldn't be!"

So back in 2015, two extremely sensitive lasers proved the Earth was squeezed by Einstein's Spacetime (matter or fabric). Just like a stress-ball is by a writer with a nearing deadline, or a diver in the deep-sea feeling the pressure of the water. Einstein's GR received a gift, as a prize for surviving till the age of 100 and becoming a Centenarian, in the form of being reconfirmed. Here, again, I won't bother you with its price-tag. But, "what is this rippling spacetime?", some of you do demand now to know. Remember the water analogy from two paragraphs ago[1]? I believe no human can come with a better one. But the use of 'water' as spacetime is just not sexy enough. Consequently, some may call it 'Foam', while others imagine 'Jelly' or 'Honey' as some spacetime-matter analogy. The *Superfluid* term and concept is gaining real favor in the hearts of many cosmologists. This will probably not-end-well, since superfluid in physics is very well defined, and they will be looking for its traits in actual spacetime. Me, I like these all, but still prefer a completely new term, keeping mine and their minds open. So, allow me to suggest – *AetherFluid*.

GR is Einstein's crown jewel, and you know, the dude sure wore some jewelry. He himself defined his Spacetime-matter

---

[1] If you don't, please, put this book or device down, now, and schedule an appointment with a neurologist.

as Aether-like, and I am not the only one out-there who's attempting to reintroduce it to science. The formalizer of Gravity prior to Albert was Newton. Even Isaac didn't like Newton's Gravity, since it suggested some far-reaching force, "materiel or immaterial" (in his words), pulling stuff together, without explaining its physical agent or medium. But it was the best theory yet, which means, it was the closest model to the observations. But Mercury's elliptical orbit was wrongly predicted by Isaac's model, and it was laying there for the taking (or for a refitting). Einstein seized it with two hands and one insubordinate mind, and theorized a model describing how Mercury will move in an AetherFluid-type-of-spacetime; and, how this space-fluid reacts back to the matter moving within it, and creating the force of gravity. That's how gravity emerges, according to Albert. Einstein knew his new model must fit where the older one didn't. So that's what he did. He assumed some fitter assumptions and wrote some new mathematics for the data to fit it. This gave birth to GR, and mainly to the *Universe Curvature Geometry* which determines the strength of the force of gravity, and with it the flow of time and the cosmic behavior of particles of light. The more massive planets or stars or blackholes are, the more they distort the near-by spacetime-matter, consequently creating some AetherFluid's 'multidimensional-slopes' that cause matter to 'fall-down'.

Einstein (very clearly) told us that gravity is an illusion, that it is entirely emergent, not at-all fundamental. Nevertheless, our leading physicists still try to find its quanta, the so-called *Gravitons*. They're searching for particles that attract stuff to more massive stuff, or, particles of the spacetime itself that exhibit some force-of-gravity. What they should be doing, is to understand what Einstein tried to tell them. Spacetime curvature creates a-kind-of 'multidimensional-slopes' and 'paths'; and, the only relevant particle they can attempt to search for (which they'll never detect), is of spacetime, of the AetherFluid

Another statement Albert made with zero caveats, was that: "Time is an illusion", which I totally agree with and am on a quest to convince you too. Just like Shlomo Barak, who also attempts to reintroduce Aether, and has his own views of what is this spacetime-matter and its geometry and dynamics, I listen with care to the words and intuitions of the most beautiful of minds. And, also like him, I read all physics' papers and books with a pinch of salt. Why like this? I ask myself like an Israeli (that I am) when in India. That's because physicists are human-beings, with careers, colleagues, self-interests, mortgages; and, as previously stated, we live in some very troubling times, when it comes to science.

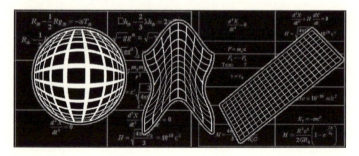

You see, and here I will risk of sounding a-bit-of-a conspiracist (which I'm certainly not), but they didn't find any Boson called Mr. Higgs, and the detection of GW advances us, very generally, just plain-nowhere. The amount of money spent on finding the Higgs just did not allow

not finding it, and the latest evidence suggests it is not an elementary particle after all, but it is made of other particles. Not to mention the fact that they can't calculate its mass. A big blunder. Also, the detection of GW did no-more than providing evidence to what was already known, even if they claim that now physicists can observe systems that are invisible, or impossible to detect by any other means. They are about to engage in the 'sport' of inventing even more un-substantiated stories of what happened this-and-that billions of years ago, in galaxies far-far-away (when blackholes spiraled and merged into one, for example). I don't believe they even believe themselves. I don't.

The case with QM and QFT is even more complicated, since this model was and still is built a-bit patchy, and way-too-much ad-hoc(y). I told you how and why they have invented the Higgs, and now I am going to tell you just a-bit more about an even more controversial and elusive piece of physics, they called – Neutrino. This one is very special. It has no charge (no +/- electric charge, thus it's neutral); it interacts with almost nothing, surely not with the Higgs (yet it was discovered it has mass); and, this ghostly particle is probably found in the universe even more than light, i.e., there are more neutrinos than photons flying all around us. This particle was invented to account for the loss of energy in several quantum processes. It was supposed to have no mass, yet it seems it does have some, and since it is not interacting with the Higgs, another mass-creation's mechanics is needed here. Told you, quite ad hoc(y) is QFT. The neutrino is more than a mystery; it is a super-blunder, an ultra-one. I believe the reason the neutrino does not receive any mainstream attention, and only very few references by our leading physicists, is because this one paints the entire Standard Model (SM) in screaming lights of incompleteness, and it certainly doesn't add to our physicists' certainty when they advocate the SM.

1950 to 99

The years that followed humanity's recovery from World War II, and (physics' triumph of) the atomic bombs that concluded it, were jam-packed with experiments tracking the theories and discoveries of the 55 years prior. Our understanding of both the cosmos and the quantum, and our technology, greatly evolved. The beginning of the 1950s was also the time when the US Central Intelligence Agency (CIA) established Project MKUltra, introducing LSD to the United States. And in the late 1950s, 'magic mushrooms' began to be eaten by 'the white man'; right after a mycologist (a mushrooms researcher), named R. Gordon Wasson (a banker no less!), traveled throughout Mexico and participated in **Shamanic Psilocybin Ceremonies**. He lived to write about it in *Life Magazine*, where **Timothy Leary** firstly read about it. Hilariously, later it was published that Gordon was on the payroll of the CIA during his travels. Yes, the US government paid for his trips.

Even more hilariously is – "How to Change Your Mind: What the New Science of Psychedelics..." by Michael Pollan (2018). Only a Western-Boomer can announce of some 'new science of psychedelics', as if it was discovered on the day he himself explored it for the first time. This science of mind-altering substances, just like the science of The Self or of Consciousness, is and has been the focus of indigenous cultures for millennia, and possibly even more. So please, let's show some respect, ok? It's completely unscientific to restart a research from scratch, without a proper review of the existing literature, even if it's not in English or not written at all. Sometimes it happens that science does not happen in the lab, and what is required is to actually talk with human-beings. I swear. Now, after this quick anecdote-break from the history (and blunders) of physics, let's get back at it.

Retrospectively, it appears that during 1950 to 99, cosmology was just slightly more of-the-focus than the quantum, and surely much more than psychedelics. I think this is true, even though it was everyday-technology, sprang from an enhanced understanding and usability of electrons, that evolved and impacted humans the most. I think it has a lot to do with the fact that we've gained the tech needed to truly probe the quantum only since the 80s, once economic growth (and boundless money printing) started kicking-in. Nevertheless, the undoubtedly biggest of all quantum disputes was settled in the 80s, the great physics' standoff between two (secular) Jews. The German-born Albert Einstein and the Dane Niels Henrik David Bohr[1]. But, for

---

1 Excuse me for the appropriation of Bohr to 'my-people', but a Jewish mother is all our religion requires.

just a while, we will focus on the cosmos and not on the men who wrestled over the nature of its particles. As told, after more than 100 years, physics still views them, the cosmos and what is inside-of-it, entirely ununited.

Most notably, the argument between the 'Steady State(rs)' and the 'Big-Bang(ers)' was settled in favor of the latter, once Alan Guth's *Inflation* was assumed. In his 1949 BBC radio series, "The Nature of the Universe", Fred Hoyle (mockingly, though he much later denied it with all his heart) called the expanding-from-a-point-universe: "The Big-Bang". He believed that the universe was 'steady-stated', as Einstein (made his biggest blunder and) initially thought. The theory and the catchy term have become mainstream, not only in astronomy, but also in society. The dominating framework to describe the universe's creation is called – *Cosmic Inflation*, and it was invented to explain the 1965 (accidental) discovery of the *Cosmic Microwave Background* (CMB). Sadly, for them, the CMB didn't support what physicists had then thought, the then accepted model of the cosmos. As a solution, in 1980 Guth (and others) have proposed an *Inflationary Big Bang* universe, as a possible explanation to the *Horizon and Flatness Problems* caused by the CMB data. Data; light; photons from a radiation that can be detected from any direction, filling the cosmos. They offered a non-linearity to the universe's initial expansion, by **Fitting a Model** to the new-found data. With this refitting they've updated their Knowledge, i.e., the Predictive Models. Yes, just like that. That is Science. Please, don't you doubt that. It is not so 'bad'. For your information – we Vedantists don't really care for good or bad, only true or false.

Why Science and Psychedelics Go Hand-In-Hand | 47

Strictly cosmically speaking, the last millennium ended with the 1998 controversial evidence for the universe's varying expansion rate (since the Big-Bang-Inflation), and the first direct evidence for a non-zero *Cosmological Constant*, Einstein's blunder that turned out just fine. Very soon more on that. Here I feel safe to say, that since 1999 the 'dawn of dominance' of Dark Energy has commenced. From then on, we are told to be presumably governed by an all-pervading and very-unpredicted force, pushing everything, especially spacetime, around. Again, physicists use terms that can easily be replaced with – 'God(s)'. We are not really sure (have-no-idea) what energy is, and certainly we're not certain of that dark-and-all-encompassing one, but, they loudly state that 'it' rules over our lives.

I will sum-up what we knew about the cosmos' creation, by 1999. Around some many human-years ago, all that exists, everything that is now in you and in the skies, was condensed to an NBA-ball-size of some tremendously hot stuff. Then, in exactly, not roughly, yes, exactly 000000…24 more 0s…000000.1 of a second, "God created the heavens and the earth" (Genesis 1:1, The Bible[1]). It happened quite

---

1   It's a great book, I highly recommend. It was very influential, no less than Bohr or Einstein (or even combined).

similarly to how an Italian chef extends and stretches a pizza dough, but really-really-fast. Like, really fast. That's how they say the universe came about; and they even give it some very-accurate time-frame – within just a 000000...24 more 0s...000000.1 of a second, it all happened. Now, seriously – how can it be taken seriously? I find it as even more flippant than the folklore-story of the spinning blackholes from the GW's detection. I am not saying-so, nor am I equipped to state otherwise, but, 'calculating' precisely how-fast it all had happened, is something that I, as a researcher, a scientist, (dare-I-say) a philosopher of statistics and knowledge; just find it hard, impossible even, to swallow. Not to mention the fact that speaking about 'time' in such an existence, is simply misleading. This I can and will easily prove in Chapter-8. Now that you know how they say the universe was created, let me tell you what they say there's inside of it, what stuff is occupying the AetherFluid.

On the quantum front, the 'weirdness' that came to be known synonymous with QM, grew even further, and much deeper. In 1957, the *Many-Worlds Interpretation* (MWI, AKA 'The Multiverse') of QM, is formulated by Hugh Everett. The MWI is just a great example of fine-math, that simply relies on utterly-ridiculous physics and philosophy of science. As mentioned, QM, the model of how little-point-like-stuff behaves and interacts, is just an 'on-average theory,' a **Probabilistic Model**. I shall dive deeper into models in general, and specifically quantum models, in the next chapter; but I must state something here, first. Physicists got totally blundered with their own thoughts, completely confusing between predictive models and (real) reality.

The Wave Function is a statistical tool that some scientists, like Sean Carrol, think is real. He even published a book advocating this. "Something Deeply Hidden: Quantum Worlds and the Emergence of Spacetime" (2019). I pity the 'Physicelebs-Multiverse(rs)', who took the place of the (failed) String-Theorists.

They believe, and it can never be even remotely confirmed, that all interactions, even on the very subatomic level, split (or branch) existence to all possible interactions' results. Anything that can happen, any outcome within the information (mathematically) 'held' in the Wave Function (WF), happens. All exists – in parallel universes. Sean even thinks that there's a *Universal Wave Function*[1]; and, that there are many, many-many, minutely different copies of you and these words, somewhere. Once a die is rolled, and while it is still rolling, the outcome is (most physicists claim) in a state of *Superposition* (many places at once, and we'll be back to this absurdity as well). And the expected-from-information result, which will be between and including 1 to 6? They all happen, all the outcomes of measurements or observations come-to-be, just not in the same world.

---

1   Again, just like Dark Energy, you can place the word 'God' there instead, and it still runs well for most humans.

The universe, you and the die, everybody and everything must split into 6 worlds, minimum, just because you have rolled a die. Indeed, some really believe that, some even build careers on it, and somehow, it's becoming accepted. That's what happens when a statistical tool, producing probabilities for events, is considered as a real-wave that either 'collapses' or 'branches'.

Contrary to the obliviousness that gave birth to the multiverse, Bell's Theorem was a virtuoso of statistics and physics. It was suggested in 1964, and tested again and again since the early 1980s. Bell is at the top of my list of past-scientists I'd love to have a beer, or even more, with. You see, and it is in the next section, but for 50 years the greatest debate between humanity's most disruptive minds, over the deepest facts of existence, was left unsettled. Einstein, clearly (no less than) despised the probabilistic nature of QM, completely disbelieving the randomness at the core of reality, that Bohr preached for. He assumed he was disproving QM when he found that Entanglement is hiding in its math. It is "*Spooky Action at A Distance!*", he and his colleagues mocked. A-bit like the 'spooky action' of Newton's Gravity, which we know was deemed-incorrect. But, here, Albert was wrong, and quite the opposite had happened. Bohr won the entire war; and not only that, randomness became an accepted aspect of reality, and the natures of space, light, and that of time, got looser and flexible. The more we put QM to the test, the more it hits us back in the face. John Stewart Bell was the first to offer a method to test one of the craziest ideas of QM, of a faster-than-light influence and synchronization between particles that once had interacted. Due to Bell, Entanglement was deemed (very) real, Einstein was declared wrong, and a new age of weirdness in QM was born. Indeed, that's how real science goes.

I shouldn't move forward without mentioning Clauss Jönsson, and his 1961 version of the *Double-Slit Experiment* (DSE). He was the first to use electrons instead of photons in the DSE, yet still he produced similar results to those of Young (who in 1801 taught us that light's nature is a wave). Jönsson confirmed that electrons also behaved according to the Wave-Particle Duality (WPD) of light's quanta. Not only photons are wave-like-stuff, all matter is like that, and even the not-so-small things. And it's even more astounding than it sounds. Soon more on that. For now, know that the WPD is the source of all the philosophical and statistical blunders of, among others, the Multiverse interpretation of the WF. It is also the reason why most physicists believe that a particle is spread all over the place, literally (so they say) being in several places at once, in a 'superposition' of probabilities, prior to it being measured. I know, the last sentence was long, even for me, but, don't you worry. By this read's end you'll get sick of hearing about WPD or the DSE. "Why?", you ask. Well, "The DSE contains the only mystery of QM" – said R. Feynman.

1900 to 49

(Now those were the days…) The third era to be told of was probably the most important for physics, and, I believe, the most impactful on modern (digital) societies. For good and bad, always going hand-in-hand, we still feel this era's ricochets. Especially during the last half-decade-or-so, with the rising of **Post-Truth** and **Populism** in pluralist-liberal democracies. And in these-very-days, since early 2020, with the COVID-19 developing to become a stress-test to the liberal-foundations of liberal-democracies. It seems like it is failing…but I slide off-topic…so let us get back to the third era in the history of physics. If we just overlook the annoying two World-Wars (which my grandfathers' Jewish-Polish family found impossible), those were some exciting days to be alive, and to 'do-science'. It was a time of such reshuffling of our understanding of reality; decades a researcher can only be jealous of those then alive and in-the-game of discovering what-is-real. It spanned from Max Planck's 1900 (out-of-desperation) mathematical-idea of *Energy Quantization*, which he didn't believe was real and initially did not endorse Einstein's use of; till the 1948 formalization of Quantum Electrodynamics (QED) and Quantum Field Theory (QFT), by a couple of men and one Feynman (who was also Jewish, by the way). It was truly a remarkable time for some great (male) thinkers, yet, not so much for women.

It was mostly the era of Albert Einstein. Special & General Relativity. His proof of the Photon of Light, to solve the *Photoelectric Effect* (his own contribution to the Quanta, for which he received the 1921 Nobel). *Brownian Motion*, that he utilized to prove that Molecules are real. As mentioned, the dude sure made some impact on the field. He was just so-damn-good in merging ideas and equations, and telling a compelling story. But, and there is a 'big-but' to mention, we can learn from his grave mistakes no-less than his major successes. His decades long quarrel with Niels Bohr, over the nature of quanta and reality as-a-whole, and his self-inflicted (biggest) cosmic-blunder, the Cosmological Constant in General Relativity (GR), are two top instances of his errors. We should always remember – even the most-open-minded can occasionally get-shut and quite-fixated become. That's what happens when **Ego** takes-over the science. Now, let's start with Max Karl Ernst Ludwig Planck, the German who suggested a simple math-hack, entirely disbelieving the reality of his idea. Still, somehow, completely by accident, he hit the nail right on its head.

It is quite easy hacking a problem with math. Accountants do similar hacks almost every quarter, adding some plug-number to an imbalanced report. Just like that, and then it works-out. Planck did something quite alike. Max built a model, meaning, he envisioned an assumption and built an equation on-top of it. His idea was that energy of light comes in packets, in quanta, in 'minimums'. It was just math, nothing more, nothing real, nothing he saw (or could ever see). He tried to solve a problem that physicists have long been obsessed with – *Black Body Radiation*. I'll skip over the problem, which is always in the nature of data that doesn't fit the accepted model, but I do wish to focus on his deed. All he did was to say something like: "If energy of light would exist only in a minimum amount, that's always multiplied by some positive integer (1,2,3...), everything will work-out just fine." And that is how the Quantum Revolution has begun, with a math-hack, yup. One cannot overlook this fact, or accept it lightly. This is how science often advances. Most of science, especially the field of *Medicine*, relies on – **Trial & Error**.

It took Albert Einstein less than five years to jump on the 'Quanta Bandwagon'. His imagination (and appetite for fame and recognition) did not allow him to overlook such useful mathematics. Essentially, he (ab)used every piece of math that seemed useful to his ideas. He was in search of the most provocative of theories, to prove or solve the most serious of problems, and find them he sure did. In a single year he unveils the theorized existence of molecules, the unthinkable reality of light's quanta, and the nature of the flow of time. For the latter, he used yet another useful 'math-only' idea, *Lorentz Transformation*, that allowed him to approximate different time's flows for objects moving in different velocities. He introduced *Time Dilation* (TD), completely rethinking Newton's ('godly' and steady-stated) Universe. TD is the phenomenon of time passing slower for observers who are moving relative to another that is at-rest. Without TD we have no GPS, without which most modern-humans get just nowhere, relaying so heavily on navigation and/or transportation apps. But, that is also due to General Relativity, to spacetime's curvature and the (so-called) *'Gravitational Force'*, that also influences time's flow. Soon enough I'll elaborate on that, and a-bit later we will dive as-deep-as-possible into – Time.

Just a paragraph here before I start with the name-dropping and their scientific contributions (and philosophical blunders). A name not so commonly mentioned, I do wish to mention here. That is the Londoner Sir Geoffrey Ingram Taylor. In 1909, he was the first to perform the (almost) single-photon-per-shot Double Slit Experiment (DSE). Later it was proven he was only 'close' (to single photons), but it still produced the same results. The basic DSE is as-easy-as-pie. There's a light source, a screen to absorb the light, and some barrier between, with, quite-obviously, two slits. When Young, in 1801, first did this experiment, the result on the screen suggested that light moves just like water, in waves that create *Interference*. He concluded, from the end-results only, what it is that had happened in-between. Remember this. So, since Young, light was considered to be of a wavy nature. But, if Einstein was right and light is just a-bunch-of Quanta, of discrete particles (he called Photons), what is interfering with what, and exactly how? They can only interfere with themselves, scientists concluded. But still, if a single-photon-per-shot is doable, as it was almost for Sir Tylor, what is that interfering with what? Consequently, the irrational idea of a Wave-Particle Duality (WPD) of photons, and later of all particles, was born. And an even crazier twin-idea, suggesting that a single photon goes through both slits simultaneously, interfering with itself as a real wave, was born a second later. The DSE's story only starts here; much fun ahead. It, the WPD that had become accepted as the nature of matter, still haunts physicists and philosophers of physics and science to this-very-day. By the time we're done here, I promise a fresh take to this ludicrousness.

Now back to the flow of history. The 1911 discovery of the *Atomic Nucleus* by Ernest Rutherford, which was another sort-of-a lab accident that followed by a brilliant intuition (crucial for science no less than math), was the actual explosion of the 'Quantum Race'. Directly it led to Niels Bohr's 1913 cool model of the atom. Once it was established that the atom is mostly empty, the quest to find a model describing it was in full-force. That was before some 'crazy-ideas' became mainstream, before physicists started believing that particles can physically, not 'in-probability-only', be in several places at-once (before measured). Bohr's model, incorporating the quantization idea, was just very fitting and accurate (approximation) to be overlooked. It was conceived to solve the minor questions of – "Why atoms don't collapse into themselves, and why anything exists at all?" I guess, Bohr to himself said: "If you can say, like Max and Albert, that things in nature come in minimum discrete packets multiplied by integers, then, perhaps, atomic *Electron Orbitals* are also, just maybe, discrete and kind-of bordered?" And just like that a model rises from thoughts. You see what Planck had caused with his mathematics? A quantized universe, a reality of point-like-particles. Which is, by-the-way, the only reality our mathematics supports. But then a World War, the first chapter (1914 - 1918), was ignited, which kept everyone just-a-bit-distracted. Well, everyone besides one stubborn Albert Einstein.

Before I'll tell you how Albert accomplished his crusade, his personal war against the late Newton, here below you'll find some names and some of their ideas that built the foundation of modern QM. So then, during the mid-1920s (and till these days), physicists started chasing their own (quantum) tails, inventing ideas and witnessing weirder and weirder results for experiments, relying on weirder and weirder interpretations of math. It came about, grosso-modo, something like that.

Louis de Broglie's 1924 *Matter Waves* made the WPD a 'reality' for all sub-atomics (now we know that for even much higher constructs of particles). In 1926, *Schrödinger's Equation* (not the *Cat*) showed how Waves evolve in time; and also, *Max Born's Rule* asserted that QM should be a probabilities-only theory, without any causal explanation. Heisenberg's 1927 *Uncertainty Principle* cemented QM *Complementarity*[1], and placed a 'limit-sign' on our knowledgeability. And, last but surely not least, *Dirac's Equation* in 1928 adds Relativity and Interactions to the Quantum, paving the way for the modern Standard Model of Particle Physics. I know it's a sin to so swiftly gloss-over these developments, and I do intend to tell of these days in the chapter on models. But, in case you do wish to deepen your understanding of how QM became so 'weird', Carlo Rovelli had just published a new one, exactly on those days. "Helgoland: Making Sense of the Quantum Revolution" (2021). A book-factory this Carlo has become.

Einstein hated all of it, literally despising the 'Quantum-Realm'[2]. By 1927 Bohr was accumulating followers and gaining vast support to his *Copenhagen Interpretation* of QM.

---

1 The principle that such very small objects have properties that cannot all be measured simultaneously.

2 A Multiverse's dimension accessible only through magic, according to Marvel Cinematic Universe's Wiki.

Bohr proposed that a Wave Function (WF) is real, describing where very-small-entities, like electrons, could be found. But, these entities didn't really exist as particles until someone had 'looked-for' them. Yes, the act of observation, an act of measurement, causes matter to exist. "(Quantum particles do-not have an) independent reality in the ordinary physical sense," in Bohr's own words. That's how deep the debate has gone. "It is wrong to think that the task of physics is to find out how nature is," said Bohr. "What we call science," Einstein firmly disagreed, "has the sole purpose of determining what is (real)." I don't think it can get much deeper than this. Bohr preached that observations create reality, meaning, that a WF is the nature of matter, and it collapses to a point-like-particle only once it's measured. In my language, Bohr believed in data, and that data does not exist before it is being obtained. Prior to a measurement, a particle is no more than just a wave of information, assigning probabilities to a 'list' of data-points that may-be created. Where the wave is higher in amplitude, the probability to find the particle there, or with this-and-that set of properties, is correspondingly higher. I couldn't agree-and-disagree more with Bohr, but Einstein completely disagreed. One must understand, and you'll find it in a couple of paragraphs, but by 1927 Einstein was known as the smartest dude on planet Earth. How could he be so profoundly wrong, so very-much not-on-point? Well, he unquestionably was, and on more than one front.

The core of the debate went as follows. Because of the DSE and the crazy result of the single-photon-per-shot; that hinted to a wavy-nature of even a single particle, interacting with itself and being a wave before it's being observed; Bohr concluded with the notion of a *Wave Function Collapse*. By all means, this is still the most widespread interpretation of

QM. Einstein, whose head was by-then deep in the cosmos, couldn't agree with the non-objectivity and inexistence of matter while it's not looked-at. He was most positively certain that the Moon was there even before humans were looking at it, in-awe or not. Stubbornly, he was fixated on the line of thought suggesting that everything in nature, on all levels and scales, had to be 'classical' just like any other 'big' object. This is what happens when you have a 'mysterious' WPD of nature; and, when physicists conclude what happens during the flight, in-between and after the two slits (in the DSE), without actually seeing what's there occurring. This is exactly what happens when philosophers, who know how to ask questions (more often than not – to a fault), are not invited to the party, nor allowed a seat at the table. Crazy ideas arise, semi-religious camps are formed, money and fame gain center stage, and scientific blunders take root and grow deep. But, that didn't stop QM from evolving into the Standard Model. After all the slanders above, I must professionally admit. For a model, it's simply the best. There's no-doubt about it.

Dirac's Equation was the first description for interactions of particles, which later received the name – Quantum Electrodynamics (QED). It still took more than 20 years for Quantum Field Theory (QFT) to come to life, and become the dominant paradigm of both physics and its philosophy, unfortunately. It produces very (on-average) accurate results, and considered – "the most successful model in the history of models". You know, results with many-many decimal 0s (zeros). During the 1940s, QED had greatly evolved with the help of the works of Richard Feynman et al. (and others). Feynman utilized the important *Principle of Least Action*[1] to formulate the QED's calculation rules, by building-on his own *Feynman Diagrams*. These were derived from his 1942 PhD thesis, using the *Path Integral* which assigns probabilities to particles' trajectories. He started by asking this question in class: "What if there are more, infinite, slits in the DSE?" Indeed, he feared none. QFT shares framework with QM, only here quanta are just excited-points-of-fields. Surely and yet again, a fine mathematical construct seems as a real physical object in the eyes of physicists. It had evolved from QED, and then had to be renormalized[2], since the original mathematical formalization of QM with particles' interactions produced just too many infinities. The energy (or mass) of the electron is just one example of an irrational infinite, before it undergoes *Renormalization*. To the rescue, came Feynman with his diagrams, that represent QFT's processes in terms of interactions of particles, alone. But, before QFT took over physics, a single photo declared the cosmos according to Einstein – a reality.

---

1   We are most-definitely going to be back to this one. You won't believe its implications on math and existence.
2   A procedure where nonsensical infinites are absorbed into measurable quantities, thus yielding finite answers.

So far in this sub-chapter of the 50 years from 1900, I told you of how QM was born and started to walk. From Planck to Einstein to Bohr, and to his disciples from then on. By 1949 QM was running like Woody the woodpecker, 'meep-meeping' through science's roads. But let's go back to Einstein now. One thing everyone can agree upon, is that Albert was the-best-ever at telling scientists what their theories (also and/or actually) state. He proved to Planck that light-quanta are real, using Max's own minimum allowed energies' math-hack; and, he explained to Maxwell that *Maxwell's Equations* logically prove the constant speed of light (photons), which without it Einstein had no new theory to offer. Also, he informed the men who developed the Lorentz Transformation that it implies to some *Relativity of Simultaneity*, and to Time (Flow) Dilation. It makes me think that this is an important characteristic of (Jewish) Albert – his Chutzpah[1]. Yes, sometimes a scientist must be able to tell his contemporaries: "All you motherfuckers are just dead wrong," or, something like: "That is what you guys are unwarily, essentially, claiming here." Indeed, again we are at a point where I must quote the witches of Dune: "Fear is the Mind-Killer".

---

1 From Yiddish – audacity, cheek, effrontery, gall, hardihood, nerve, and temerity (are needed for science as well).

We return to the Cosmos. Back in 1906, it went something like this. Albert was on a quest to shock the world. Bit by bit, so he did. Molecules, photons, and the constant speed of light that resulted in a varying flow of time, and, the equivalence of energy and mass[1]. But all these weren't enough. What else could he attack? Newtonian Gravity was then the target. General Relativity (GR) is just another glorified name for Einstein's Gravity or Albert's Universe. As mentioned a few pages back, in GR space and time merge into one fabric called spacetime, that within it everything moves; and the matter that moves or spins within it, curves the AetherFluid in return. Its punchline is quite powerful. This curvature of the AetherFluid is what 'creates' gravity, these some-kind-of 'multidimensional-slopes', or a-kind-of 'paths', between the Moon and the Earth or the apple[2] on the tree and the grass of the land. This is the same force, just on different scales, more-or-less. That was Albert's imagination, which needed no-psychedelics. Somewhere in the back of your mind, I hope you remember that he also told us that gravity, just like time, is no-more-than an illusion. I promise it will be reconciled, and even unified.

In 1915 Einstein repainted the universe, but with a slight mistake, with just a little blunder that grew into a rolling-down-snowball. For decades he danced and spiraled with his "greatest blunder" (his own words). GR's math predicted, mistakenly[3] he saw, a non-steady-state cosmos, which then

---

1 This equivalence has been philosophically, metaphysically, and mathematically, assumed by him.

2 For the falling-apple it is slightly dissimilar, since the Earth has its own set of 'quantum-rules'.

3 He simply miscalculated his math, missed the fact that it did support some static universe.

was-just-not what physicists had thought. I bet you already know what's the common course of action for a physicist with an unfitting model to data, right? Yes, I know you know. He refitted it with some additional math to lower its bias. Just like that, the Cosmological Constant (CC), just a math-term, came to life. The CC in GR's equation was meant to 'steady state' the universe, but, you should by-now also know; the 'Big Bang(ers)' eventually won the war over the cosmos, and since the 1980s physicists believe in some Cosmic Inflation, because it fits the CMB data. But, much before, since 1923, scientists were aware that the universe expands. It was Edwin Hubble, the man with the telescope's name, who reported the many moving galaxies.

By 1999's end, Einstein's biggest-blunder made a grand comeback, only this time it was no longer a blunder. The controversial evidence for the universe's varying expansion rate 'vindicated' his first full-equation of GR, including the CC term he so-hated. Lucky him, it turned-out that a piece of math, such as the CC, is needed to solve another miss-fitting of a model. I just love science, it is just so-very *Human*. Told you, everything turned out just fine for Albert Einstein by 1999. Nowadays, the CC is our best math description for Dark Energy (DE). And DE? It's as close as it gets to a physicists' God. The first real validation of GR, very-generally and with no relation to the CC, came in the midst of WWI. In 1919 there was an expected eclipse, and the new theory in town predicted that stars from behind the Sun will become visible. Since the Sun is so massive, its 'gravity' should bend spacetime so much, that photons won't reach the Earth in straight-lines, but in bended-paths, due to a curved-AetherFluid. By the end of the 1920s, GR was on solid footing. Eclipses continued to corroborate Einstein's predictions, and it was very sexy. That is how and why Albert's family-name had become synonymous with – Genius.

From Newton to the Electron

(Now those were-not the days...) In 1687, a British-man establishes the field of physics. He builds on *Galileo Galilei's Inertia*, who himself grew-up in the tradition of **Pythagoras**[1] (570 BC). The Laws of Motion and Gravitation are formulated by Sir Isaac, and **Movement** is cemented as physics itself. Up till these days, mass is an energy that pushes-around or binds things up. Even the AetherFluid itself moves by the will and energy of, well, Dark Energy. Newton is also the creator, or discoverer if you'd like, of the amazing *Calculus*. The *Derivative*, and its mirror image the *Integral*, are both 'made of calculus'[2]. I'm going to use one of them in the chapter after the next, to offer a simplified view of particles. I'll be displaying the built-in quantization in/of math. Just a quick math-talk we'll then have.

---

[1] One can easily view Pythagoras as some-kind-of an obsessed-with-nature Shaman.

[2] Before someone asks for a grant; there is no Quanta – no 'Calculuson' – to Calculus.

Now it's the early 1860s. A Scottish physicist (yes, that kingdom again), James Clerk Maxwell, demonstrates that the Electric and Magnetic fields are of a single one, that travels through space in the form of a wave, at a light's speed. He declares that *Electricity*, *Magnetism* and Light, are all manifestations of an *Electromagnetic Force* (you should know what's this force's particle[1]). Maxwell wrote his first scientific paper at age 14, but later he 'did-science' similarly to Einstein. He collected and unified laws that were originally derived by Carl Friedrich Gauss, Michael Faraday[2] and others, all into one consistent theory. Einstein later just added to Maxwell's Equations by proving that his wave is made of Planck's Quanta. And hallo Quantum. It won't be long before they find the electron. In 1897, another (yes another) British-man, Joseph John (JJ) Thomson, discovers electrons. JJ declares the birth of the first sub-atomic, without which your phone will fall through your hand. Essentially, there is no phone without electrons, since all Electronics (Music including) relay on the usability of electrons. The electron, unlike his elusive cousin, the ghost-neutrino, is one of men's best friends. Almost like the dog. Almost.

---

1   If Light is a manifestation of this force, then its particle can only be the Photon.

2   *Faraday's Lines* can clearly be the first modern idea relating to the fields of QFT.

Only a single woman is carved in the pages of physics' history books; besides Mileva Marić-Einstein, the-wife-of. Marie Curie, who's considered to this day to be the Mother of Modern Physics, discovered in 1898 the elements of *Polonium* and *Radium*. For which she received the third Physics' Nobel Prize in 1903, and paid with her life in 1934 (aplastic anemia). Days of extremely unsafe science, very-not-recommended. Can we even begin to imagine the world and societies we could have been living in, if only women were of more influence? No one can find any excuse for the fact that only a single woman is mentioned in this chapter. Only **Elements of Fascism** allow such a social environment. Quite often, probably since manhood is considered more egocentric, psychedelics and their experiences are viewed as more feminine, emphasizing 'right-side-of-the-brain' thinking. Nowadays it is viewed as an over-simplistic *Model of the Mind*, but, the intuition is still valid. Psychedelics' consumption allows the mind to detach itself, if only for just a short while, from its everyday worries and fixations, and also its ego(maniac) aspirations and obsessions. Then, one can properly do-science, either of Physics or of the Soul, with an open-mind. For physics and science to truly advance – that's what's now needed.

I wish to conclude this chapter with *Statistical Mechanics*, which is one of the pillars of modern physics and its philosophy. Personally, **Statistical (Machine) Learning**, i.e., building models by data-fitting, is what I do for a living, basically. Statistical Mechanics is describing how 'big' observations (like temperature or pressure) are related to 'small' discrete events that fluctuate around an average (or *Mean*). *Ludwig Eduard Boltzmann* was ahead-of-the-curve, when it comes to models and curves, or understanding matter and atoms. He realized some ideas, from Brownian Motion to the Uncertainty Principle, much-prior to them becoming physics' laws and common knowledge. He was also quite naturally 'psychedelic', just in a bad way – of the schizophrenic and suicidal (1906) type. Such things can also happen to those who carelessly-trip, and to those who mix drugs or deal with dirty chemicals. So please, I'm begging here, if you do consume psychedelics, bidding to have some of the abilities Boltzmann naturally possessed, do so with extreme care. You simply have no spare-brain. Maybe Elon Musk has one, but you probably don't.

Now I'm going to take a risk, yet again and certainly not for the last time here, and free-write my own true-mind. It is prearrangement, not some random-evolution, that had made us 'averages seeing creatures', animals that perceive matter as aggregations of trillions of fluctuations. It had fitted humans best, we couldn't have survived otherwise, or, we just would have been entirely different creatures any-other way. It is also this clear aimed-evolution that had 'placed' receptors in us, that handle psychedelics the way they can. On such high-levels, there are zero-coincidences, none-whatsoever. The complete banning of psychedelics seems immoral, under this view of what we are, and our nature. I was sure that it is just a matter of ages, and politicizing science was a thing of the past. The same way that homosexuality used to be forbidden, while now the LGBTQNIAPK[1] is, well, entirely out-of-the-closet. But, I was wrong. Facts are again – under an assault.

---

[1] Lesbian, Gay, Bisexual, Transgender, Queer, Non-Binary, Intersexual, Asexual, Pansexual & Polygamous, Kink.

Chapter 3

On Models and Averages

Together we learn. Data. Knowledge. Information.

The word "**Model**" was mentioned fairly-a-lot. As promised, from this chapter on I will deepen into and build upon what was glossed over, hinted to, or spotlighted in the two previous chapters. I wish to engage this, at first, with the natures of Models, Information, Data, and Knowledge. These subjects can be approached with high-words and loaded-terms, such as *Ontology* and *Decoherence*, but, there's another path to be taken here. David Orrell, with his (utterly unneeded) book – "Quantum Economics: The New Science of Money" (2018), did, in my opinion, the most horrible deed a popular-science-writer can ever perform. Orrell did no-more than just complicate things. He used the 'quantum language' to redefine terms and process that economics (that is – economists) had already found words and equations and clear descriptions for. He rode the 'quantum-buzzwords-wave' into financials, without understanding he is doing an unscientific manipulation. Having said that, I'm going to do quite a similar maneuver, just from the opposite side, and speak about models of physics as if we are learning **Introduction to Econometrics**[1].

---

1   Statistical methods using data to develop or test theories in economics or finance.

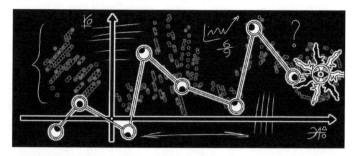

Let's build a model, shall we? Say we are in a room, only you and I, with no windows and only a single door through which someone is expected to join us. I ask you: What is going to be the height of the next person coming-in? For simplicity, we'll assume it is a healthy adult, aged 22.5; and, that outside the room we have exactly 1000 representatives from each country; and, the citizen to join us is chosen randomly. How would you answer this? What Model do you have in mind? Guessing is not allowed; a reasoning must be provided. Currently, just to begin with, no additional data is available. Meaning, that's the only knowledge you have – the information known about the next creature who's about to enter the room is that it's a healthy-adult-human. Now, what is your answer?

"Boaz," you begin, "my friend," now you soften, "I have zero chance of getting this one right, with such a wide range of numbers, from short women under 155 centimeters up to (NBA) men above 2.2 meters." You're a smart cookie!, I answer back, since you don't fall into my trap. So please give me, I adjust, a number you think will be the closest one; the number you predict that the *Error*, the difference from the actual person's height, will be the smallest. Now you really need to think-some, utilize some-kind-of model. You do think-a-bit, and ask me to google for two numbers; wanting more knowledge before you answer, as you should. I allow that, without asking for which ones. After some googling you come back to me with your answer. "1.67 meters," you proudly reply. I ask for your reasoning, wish to understand the logic. "All I did," you are explaining, "was to find the worldwide-average-heights of males (~173cm) and females (~161cm), separately, and then I have just averaged them; since the probability of the next person joining us, being either male or female, is 50%-50%." And voilà, a model outputs a number. Now that's thinking! I am impressed by your reasonable methodology. Good on you, dear-reader.

A Model

Have no doubt, that is a model. Indeed, a very basic one. But yes, the Average (of a Data sample), which is an *Estimator* to the whole population's Mean, is a valid model. How is a model evaluated? Why QM's models, the Standard Model (SM) that's derived from Quantum Field Theory (QFT), is considered – "the most successful model in history"? Because predictions made by it are (on-average-only) the closest to observations, i.e., to data-points obtained in real-life measurements. Unlike the model we have recently built, where I've told you will be evaluated in a one-time-event,

QM's models are evaluated by summarizing the errors of trillions of (analogically-speaking) humans (if that many were available) joining the room, and not not just by the next one coming. A model is evaluated by its *Predictive Power*, by averaging its errors from real observations; and, this is only the case with models of the 'small', models of wave-and-point-particles' semi-random mechanics. We'll get back to it in the coming chapter, to this concept and reality of 'point-likes'. For now, we are still focusing on models, their nature and types, and use-cases. Bear with me.

Unlike QM's models, that can't say exactly where an electron will be found, only the probability of it being somewhere (very-well defined); models of the 'big', of classical objects, should always be correct, with every error being equal to exactly zero, and not just the aggregated-average of errors. If a classics' model doesn't perfectly describe the path of a snooker-ball, a star or a planet (as was with Newton's Gravity and Mercury), it's just deemed as wrong, or, at least, as not-the-best-approximation. That's how General Relativity was born, to explain Mercury's path; and that's also why additional terms were added to it, for Dark Energy (Cosmological Constant) and Dark Matter. Precision is sexy.

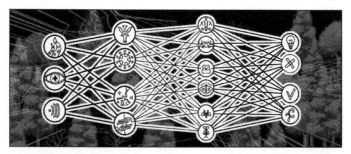

Very generally speaking, there are two means for models' building. One is based only on data. Applying algorithms that utilize statistical elements and methods that 'learn', i.e., discover knowledge and patterns by fitting models to data. Currently, the more powerful of these are variations of *Neural Networks* and *Decision Trees*. Sometimes it is (ignorantly) called – 'AI'. Please, don't ask me why. The other method to build a model is much-more mathematically-heavy and theoretically-based. Constructing pure-math-equations that are built upon theoretical assumptions, with no use of trees or networks. These equations include *Constants of Physics*, that, very-so-often, hold much of the assumptions (and the lack of knowledge) themselves. That's how Sir Newton wrote his gravity's (approximation) function. To his credit, he knew it wasn't real, he didn't believe its physicality, yet still it was taught for a couple centuries and was the-one-to-beat.

Then Einstein came with a better approximation, by relying on more-correct assumptions about the nature of time and the matter of space. However, just another approximation is this one, as cosmologists now realize. That's why Dark Matter came into our lives, in an attempt to refine and rethink Einstein's (AetherFluid or) spacetime-fabric. Images of how galaxies spiral and held together pushed physicists to believe there's some cosmic-super-glue, which doesn't interact with light. A blunder influencing theories and research of the quantum as well. That's also the main driver behind where physicists are currently focusing on – the AetherFluid. As I've mentioned, there they search, with no luck so far, for the non-existing particle of gravity (that they never saw but already have a name for).

Let's get back to that room where we sit in, staring at the door, waiting for an Earther to join. Firstly, you have used the prediction power of the naïve-average, an almost 'naked average' of the average of heights of men and women worldwide. Now I continue our model building and ask: What *Variable* would you like to know, in order to have a more accurate model, to be more precise in your next answer? You go-on-thinking, yourself asking, and rightfully so, what data will be of the most importance – will add the most to the predictive power of our model. Indecisively you answer: "I'm undecided between the Country or Gender I wish to know, of that soon-to-enter-the-room human. But, if I do have to pick just one, I will choose the gender," you finally decide. I know why you chose this one. You, just like nature, are lazy. You have already obtained the global-averages of females and males worldwide. Adding this single variable, this piece of data, will allow you to simply choose one of the numbers you have already googled, not needing to ask for permission to google all the countries' average heights. I like your (lazy) answer as much as I like nature, but it's not the right one. Knowing the country, and the average height of its citizens, of course, will allow you a more accurate prediction.

So far, we've built some-type-of a *Regression* model, i.e., a model that its output is a continuous number. Its information's space is that the actual height will be between ~150 centimeters to ~2.3 meters. Now let's make this a *Classification* problem, with a probability output, by rephrasing the question. We will start with a *Binary-Classification* problem, a model producing a single probability of a discrete event. The reshaped question I ask you is this: What is the **Probability** that the next to join us will be above the average height of 1.67m? And I add: What additional data you now wish to know, to get it fairly accurate? This time, I hint, some healthy laziness and just a-bit of knowledge will probably serve you well. You get the memo, and ask again to know the gender of the joining person. But what more knowledge was I clueing to? If one has a simple table of the average height of men and women per country, it becomes quite possible to produce an answer. You google and find an Excel file with the desired table (knowledge), and I provide the data the next person to join us, is – a she. Now, what is the probability that her height is above 1.67m? Following some thinking and spreadsheet munging, you answer: "I'd say 16%," corresponding to the percent of countries that the average height of their female-citizens is above 167cm. If we will let 1000 humans, women in this case, enter the room, probably, roughly, around 160 of them will be higher than the average height of a healthy female. I can only hope it's clear as most Tel-Aviv-Yafo days, that a probability has zero-meaning, it's meaningless, for a one-time-event and as an attribute of a single-observation (or fact).

In this paragraph I'll present the *Multi-Class-Model* case, where I ask you: What is the probability that the joining-person is exactly between 1.64 to 1.7 meters, no-more-no-less? I hope you see the three regions, equivalent to three probabilities (that'll sum-up to – 1). Now, if you are clever as I think you are, you won't give me an answer without demanding to know both the person's country and gender, wishing to produce the most accurate probability. That's the nature and reality of models with variables. More precise probabilities, or less-errored numbers, require more variables and/or (much) more data. Together, adding knowledge for a better model. That is the case with the models I build. Pieces of code that tell what's the probability of a user to in-app-purchase; or how many ads the-same user will endure before she uninstalls the app, if the first model tells me she will never buy any of those virtual coins I offer her (at an amazing sale!). I can never tell my clients anything about the next user to use their digital product. I can only suggest, that if lots-of people will play or bet, the probabilities I produce will mean something. In the 'quantum-world', one observation is meaningless, both for learning and evaluation. On the other hand, or model, for a 'big-world' theory, one good hit might be all you need.

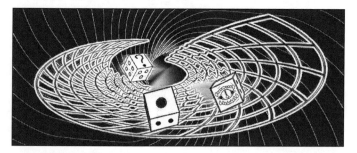

Before we'll explore the Wave Function (WF – the first real QM model), including some history, a quick one I wish to ask. If we roll a fair-die 6000 times, how would you predict that the results, the data, will be distributed among the 1|2|3|4|5|6, if plotted on a graph? Quite evenly dense, right? Uniformly distributed, like a flat-wave-line. The WF is not that different in its statistical and philosophical nature. It is just more, well, wavy.

Wave Function

By November 1926, Werner Heisenberg (with his Uncertainty-Principle), Max Born (with his QM's probability-interpretation) and Jordan Pascal (with his soon-to-be-Nazi-views), had produced the *Matrix Formulation* of QM. It is a powerful formal approach to make quantum predictions, still broadly used in nuclear physics and for engineering. It focuses solely on experimental observations, on the data. Unlike them, Erwin Schrödinger started from a definite picture of the underlying physics of nature, when he formulated QM. He thought that (Louis de Broglie's) Matter Waves, were just like Maxwell's electromagnetic waves of Photons. He was able to produce a formulation, a model, for particles' locations, in a mathematical language that was more popular than that of the matrix. Bohr (the then 'Quanta Guru' and modeler of the atom) arranged for Schrödinger to visit Copenhagen. There, he and Heisenberg held lengthy discussions in an attempt to convince him, that while his wave-mechanics formalism was pretty-valid, the naïve picture he imagined behind it was not-so-much. Bohr argued that many aspects of quantum waves' behavior, particularly the phenomena in the WF of probability and uncertainty, were not consistent with Erwin's simple ideas. Erwin viewed particles as being some-kind-of 'wave packets', and not those point-like-things-stuff. Schrödinger was not really convinced there were any interpretational problems with his views. Nonetheless, he finally consented, but retained a dislike for the *Copenhagen Interpterion* of QM for the rest of his life. His wave mechanics had to go forward without any underlying picture of what lay-there, under-the-hood.

"But what is this 'Copenhagen Interpterion' (CI)?", now you wonder. The CI, born in 1927 and named by Heisenberg in 1955, is a 'data-only' approach to QM, as I read it. Niels Bohr focused on the Wave-Particle Duality (WPD). He emphasized the oneness of the 'measured' and 'measurements' themselves, insisting that these could not be analyzed separately. He claimed that a measurement 'collapses' the wave-particle-something into a point-like-matter, with its few physical properties. His philosophy – that two seemingly contradictory descriptions together characterized the same phenomenon – was and is still widely promoted as an important part of QM. As mentioned before, Einstein detested this view. At first, it was the randomness at its core. The fact that this model produces only 'on-average' accurate results. Believing that randomness is nature-intrinsic, was just unbearable. Not long before, Albert was successful in describing Mercury's path around the Sun, with great precision and no uncertainty or randomness. How can science accept that? Do you think an astronaut would go on a spaceship, thinking there is randomness in calculating its path to Mars? Why should randomness be so-fundamental on the level of elementary particles, only; but not in higher constructs of them, such as rifle-bullets or space-rockets? Moreover, why a measurement plays any role in what is being measured? Things are just things, always, and anyways, aren't they? Well, apparently, they are not.

By 1927, Paul Dirac was able to derive a much-more general quantum formalism, developing a relativistic WF by introducing four complex wave functions. The extra wave functions correspond to additional variables (just like gender or country), which can be related to other properties of the electron, besides position. He showed that the wave and matrix mechanics could be derived from his own *Generalization*. Physicists were just less accustomed to matrixes, therefore the wave mechanics survived. The problem of the two competing formalisms had been resolved, but still the question of their interpretation remained to-this-very-day. Einstein et al., a few years later, tried to attack QM from another angle. They've found in the WF an implication to some 'Entanglement' (coined later by Schrödinger in 1935). They found a 'spooky' non-locality of the synchronization between even-far-away particles. To say 'non-locality', in this context, means something in the spirit of 'magical', more-or-less. They were sure this will be the end of the WF, and QM. The opposite had happened, as you know. It remained an open question till the 80s, till Bell's Theorem was tested. Then, this entanglement, a faster than light influence between remote entities, had received reality's approval. Yes, that is how nature operates.

Is the WF real? Is there really a wave-like-single-particle, that doesn't exist before it is interacted with, when then it becomes a point-like-thing? Is there a 'superpositioned' particle, going through both slits in the Double Slit Experiment? And, is this hazy-wave (sometimes called "cloud") of probabilities, which is never 100% correct but very accurate on-average, really real? To that David Joseph Bohm would probably (near his death in 1982) answer: "Elementary particles are systems of complicated internal structure, acting essentially as amplifiers of information, contained in a quantum wave." That's limitedly helpful, right? This view relates to his later theory, the – *'Implicate' and 'Explicate' Order*. But, before, he was the first to offer a deterministic interpretation to QM, known today as the – *Pilot Wave* (or *De Broglie–Bohm*) *Model*. "Hi," you stop me again, "who is this Bohm"?

Hidden Variables

Not to be confused with Max Born, who was also a Jew (a German born 35 years before); David Bohm was an American, born in 1917 to an immigrant family; who thoughtfully pursued a unified theory for the classic and quantum, and for matter and consciousness. In his models and metaphysics, there's no principle of uncertainty to reality, only a lack of knowledge about the initial position and state of a particle, for example, or, what other variables should be in the model. He called it – *Hidden Variables* (HV). Bohm was also a researcher of consciousness, holding views that deeply resonate with perspectives as diverse as those of Shamanism and Vedanta. Jiddu Krishnamurti was a great partner and influence on him, in this regard. If I need to reference a physicist who I think stands out from the crowd and can be a model-figure for next-generations' scientists – Bohm is that.

He proposed that what we normally think of as a point-like-particle, is a temporary localized pulse emerging from a larger field. Quite similar to some vortex momentarily forming from the dynamic-flowing of a stream. I just can't agree more. Soon you'll see. Bohm's work of the 1950s, some of it was conducted in North Israel, where he met his wife Sarah, was influential to Bell's Theorem of 1964. Later in his life, once he had opened his mind outside the measurements-only physics of inertia (and into human-experiences), he stated that his HV models and concepts were aimed to make-a-point, scientifically and philosophically, rather than being practical. He just wanted to prove that not everything might be known. What is outside a model, is simply – hidden to us.

On the one hand, he was more modest than the Copenhagenists; on the other, with HV he was also telling stories of what particles are doing before they are measured. Unfortunately, and as always, our physicists are goofy when it comes to words and expressing their minds and thoughts. On this regard, and only this, Bohm was like most physicists. Here is another example of how heavy he himself expressed (1980): "The new form of insight can perhaps best be called **Undivided Wholeness in Flowing Movement**. This view implies that flow is in some sense prior to that of the 'things' that can be seen to form and dissolve in this flow." I rest my case (your honor). If they'd ask me how to better name HV, I would advise – Dark Variables. Then, at least, they'd be consistent.

Remember the model we've just jointly compiled, when I asked you to pick just the age or country variable, but not both? Analogically speaking, the unchosen one was just some HV. That is why predictions are not accurate-per-observation, per-measurement, but only across (the average of errors of) many data-points. HV in QM, as a term, has a double use. It's the same case for the word 'dark' in cosmology, used both for matter we can't see (that's assumed to hold galaxies together) and energy we can't explain (trusted with the universe's varying expansion rate since the Big-Bang-Inflation). There are two 'darks' on the cosmic-scale. in QM, when one uses the term HV, it's either to explain the inaccuracy, the non-meaning even, of QM's probabilistic nature of a single predictions; or, for describing entanglement's 'magic', that physically no-human has any idea what its mechanics is. Way-too-many, including formidable physicists, mix-&-match and confuse between these two meanings of the same term. So, for once and for all, let it be said. 'Hidden' in QM means the same as 'Dark' in GR, lack of knowledge; and Bell's Theorem proved Entanglement, yet said just nothing about Hidden (or Dark) Variables.

The Case of $\sqrt{-1}$

In the next chapter, where I'll claim that everything is spherical and made of some kind of 'least-actions' (or interactions), we will have to chat-some-math. Just a-bit, promise you that. But in this section, in the context of the WF and QM, another useful math-tool, one that tidies equations, must be mentioned. This one, the way I connect the dots (or points), is in direct relation to both HV and our cyclical-rotating existence. Not only that the WF is a statistical-math-model producing on-average-probabilities of future measurements, the WF is also defined as a *Complex Function* that includes an 'imaginary' part to it. That 'imaginary' is the square-root of minus-one ($\sqrt{-1}$). Every 12-year-old-teenager should know the square-root of a negative-number is mathematically errored, since no self-multiplication of any number produces a negative one. That is, besides the 'imaginaries'. When these are combined with the good-old-normal-numbers, complex expressions are created, representing *Complex Planes*. If one does wish to visualize the role of $\sqrt{-1}$, I advise to hold a picture of a (bed's) spring in mind. Imagine you drive a bicycle on some plain-plane-road, only that in this type of a 'complex-movement' you are not always remaining on the flat surface. Sometimes you are above, sometimes below, each time in a different angle, spiraling your way while moving forward. You basically drive around some straight-line that is located inside your spring-like-trajectory.

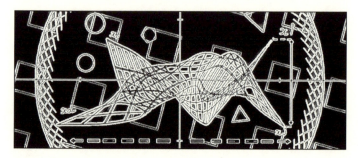

That's what this math-procedure allows – a representation of cycles and rotations in a spring-like-path. It is used as a simplification-math-tool in many fields, including engineering and physics of the classical objects, and has a direct-simplified-relation to *Trigonometric Functions* (*Sin* & *Cos*), representing waves of varying kinds. However, the situation in QM is fairly different, since the *Imaginary Number* (denoted in functions – i) appears explicitly in its most fundamental equations. There, these numbers are combined with real ones, together making 'complex numbers'. Thus, the WF is a complex function. Currently there's a lively debate regarding the real necessity (and reality) of the WF's complexity. I read this as analogous to Einstein and Bohr's squabble over the reality of entanglement and uncertainty. We have some mathematics in some model, and geniuses fight for decades over the question – "Is it real?" If you ask me (and since you are here I can only guess that you do), the scenario here is very similar to Newton's formula of gravitation. His model, which had zero-reality and just 3 variables (two objects' mass, their distance, plus the *Gravitational Constant*), approximated some much more complex physics, which now we call – GR. Today we can safely conclude that even Einstein's GR is an approximation-only model, since we have already been required to add some math for 'dark' matter (in an attempt to explain galaxies' structure, for example) and energy (cluelessly adding-back a clear math-mistake in an attempt to explain a mysterious force that cosmologists believe governs the universe's expansion).

Why was it important to mention? Why the math and fundamental components of QM must be investigated till clarified? Because, I feel comfortable to say, math will never explain existence nor consciousness, and measurements will always be limited, subject to HV and some lack of knowledge or data or information. The case of $\sqrt{-1}$ is an evidence supporting the validity of this thought of mine. Only once we'll come-to-grasp with (real) reality, on all its dimensions and levels and variables (finding the hidden-ones needed for QM's models), we may start worshiping the equations and mathematics of our physicists. Poetry, storytelling and philosophy, have a higher probability (according to my model) in succeeding to describe the complexity of reality. David Bohm and Gottfried Leibniz have realized that, and should be taken as example-figures for the type of science that is needed now. They've also understood, that if a physics' metaphysics is immoral, or does not (firstly) serve the spread of equality within our species, it is inappropriate.

Chapter 4

The Matter We Are

Quantization will be seen.
A-form-of a derivative.

As much as I've tried, it's unavoidable to have a mini-math-talk. Not to be confused with 'meth-talk', which is the overall language and atmosphere of shows like "Breaking Bad", and the reason why I haven't found in my heart fond for this so-successful tv-series. Meth is not a psychedelic nor a drug. It is poison. As poisonous (and successful) as OxyContin.

Just had to say that, now back to math. In plain words, I hope, I'll tell about some part of it, of Newton's Calculus. It'll be useful, since the Leibniz–Newton Controversy will also pop-up as a quick anecdote in the History of Math, and Leibniz should always be welcomed at any discussion about nature. Gottfried Wilhelm (von) Leibniz, born in 1646 in Leipzig, was, likewise David Bohm, just my-kind-of-scientist. He simply loved metaphysics and philosophy, and fine-philosophers by proxy. We'll get back to him when it's time to dive deeper into forces and energies. Leibniz was a true genius, a superb philosopher of nature. But for now, let's please focus on the (useful) Derivative, which Newton and Leibniz were indirectly fighting over who was the first to formalize. I do not presume to be the judge of that tiff, but, if I could, I'd tell them both this kind of credit-fight is virtually of no significance. History tells that great ideas nearly always arise by more than a single person at a time.

It is almost possibly-imagined that ideas are 'up-in-the-air', somehow, before they are 'captured'. One of those great ideas, that we will explore and (ab)use as an analogy to some physics' fundamentality, is – the derivative. To do so, as you have probably figured out, we must start with a function. Let's call it, the 'Original-Function'. No worries, as assured, no equations will be written here besides Einstein's. Quite like the regression and probability models we have jointly constructed, made of words in our minds, we'll imagine two equations, and build ideas and notions of quantization, and least-actions, on-top-of them. Every person who had ever played on their ('smart') phone or tablet a (Casual) Freemium-Game, probably while 'number-2ing', will be familiar with the first dynamics I will paint. The second will surely be clear as the skies of a sunny day, although it's more complex. That is since every adult, most probably, took a ride in an accelerating and slowing-down car at least once in their lives.

Derivatives

The story goes like this. You have downloaded an app. Let's call this 'time-zero' because we describe activities only from this-time-on. It's a cute game, in the genre you like. The game-mechanics relies on some 'virtual coin' called – Q. We'll also assume, that in every second from time-0 you receive or spend some quantity of Qs. Our function, this simple equation, only describes exactly how many Qs you own in every second since installation. It goes up when you don't play and slides down when you use the app. All depending on your in-app-activity, that lowers your Qs, and the number of Qs the **Game-Economist** has decided to grant you, in-an-attempt to lure you back to open the app. For example, at time-86400, exactly 24 hours after time-0, you had 975 Qs in your virtual-wallet. Simple, right? Now that we hold an equation in-mind, that receives a positive number of seconds and returns the quantity of Qs you then had, we can imagine and talk about its derivative.

The derivative, as its name implies, is just another function that is derived-from the original equation by the action of – *Differentiation*. This math-act is of no importance, only its result. Once the Derivative Function (DF) is obtained, it can be used to output what was the change-in-Qs at that-very-second (since time-0). That's a derivative – the change (or slope or delta) in the original function, at a very-specific-point. I hope, that by now you already think of an elementary particle when you read or hear the word – "point". Back at our simple example, let's imagine that the value of the DF at time-86400 equals 17, i.e., exactly 24 hours after time-0, while you were not playing, you've received 17 Qs to your in-app-virtual-wallet. How simple was that, right? There's a DF to the Original Function (OF), that outputs exactly how many coins you got or spent, based on what happens in your app-wallet, every (round) second since time-0. Between whole-seconds the DF is simply undefined, by-definition. That's how it works, every second your wallet is updated, and the added or subtracted Qs' value is the DF. By the way, if one preforms the math-act of *Integration* on the DF, basically summing-up the DF values from time-0 up-to a desired point-in-time, one attains-back the OF. Calculus is the "fun" in functions.

That was easier than pie, right? But why? Because the problem was, to-begin-with, so-very-much-quantized. The scenario I have illustrated had a tempo. The value of Qs in your virtual-wallet is only changing per-second, and nothing happens in-between. Now please imagine the OF is for the distance you have driven since you got into your new Tesla and started the auto-pilot. Of course, at time-zero. Quite dissimilarly to the virtual-wallet in your-favorite-genre-app-game, that's updated every second; in your car you move continuously, smoothly, like everything in nature. For a moving car there are parts-of-seconds, and-parts-of-parts-of; and in each such a 'quantum of time' the value of the DF is different, since no car moves in a perfectly steady velocity. A car always either accelerates or slows-down. Every micro-and-nanosecond, all the time. So, what's a 'point' in this case? And how can such a point have a slope (of the tangent line at-that-point), if a slope is, by definition, the difference of Y (coins-in-wallet or distance) with respect to a change in X (time)? We ought to conclude that a 'point' signifies something 'complex'. "But, what is this complexity?", you flex your mind-muscles. It is – **Event of an Action**.

Action at a Point

This next part might be considered as an over-simplification, and it probably is; but, that's why we are here, let's be real. If I would not be illustrating it simple-yet-making-a-point (points are just everywhere), it would be of interest to none. A physics' idea and term, which (you should guess) has some-kind-of double meaning, must be mentioned in the context of exploring this point of a minimal-event-at-a-point. The *Least Action Principle* is of great importance and significance to the current section. This one dates back to the inception of the field, since it allowed a definition of movement. And inertia is at physics' core, as you by-now-know. It was reimagined by the wild Richard Feynman, all within *Wheeler-Feynman's Electrodynamic*. It is the backbone of the interesting *Quantum Handshake* (or *Transactional*) interpretation of QM. As an economist, a metaphysics grounded on the concept of transactions, of attributes 'becoming' in interactions[1] only, is very appealing to me. More often than not, it is sub-easy to get lost in this one, due to the dual-meaning and the inconsistent use of the term – Least Action Principle (LAP). But its connection to Newton's Laws of Motion is clear, so I'll better start there.

In its most outright meaning, the LAP states that everything in nature is lazy. As lazy as anything can ever be. Nature is probably lazier than I; but most-surely less than you, since you're here, engaging in some non-lazy-reading. A 'straight line' between two distant points in the universe is always the shortest path, the laziest route, a photon of light or a

---

1 In (micro-)economics the demand and supply have limited meaning, like the WF 'at-rest', unless interacted.

planet roaming around a star will travel in (or on). Now, how 'laziness' is defined? Least energy spent – it is that. Simple, yet hinting at some oracle-ability of particles and matter. Somehow, particles (and moons) have found a way to 'smell' the best trajectory, to know a picosecond beforehand what action should next be taken, in order to stay lazy. All to minimize the energy spent to get there. Any other path than the laziest and least energetic one, will require additional 'investment' of force in the action.

This should feel and sound as common-sense. We humans do it constantly, taking the shortest-path-possible when planning a flight, for example. A quick reminder here, that in a curved-universe, as our existence is very-much-so, the quickest path may sometimes be a seemingly indirect one. Just as a plane from Tel-Aviv-Yafo (TLV) to Bhuntar (KUU) in Kullu District (North of India), although being on effectively the same earth-longitude-line, will start by flying north before it'll curve-back down-south. That's also the nature of photons' trajectories into our eyes from a very-far-away star, from one of those glittering diamonds you see in a clear night-skies. Soon we'll also need to see what exactly *Seeing* is, don't you go-assuming I've forgotten to define this. To summarize, photons travel as waves in (or within) Aetherfluid's paths (or roads), called – *Geodesics*.

That was one kind of LAP, nature's laziness, always doing the least-energetic-action possible. Now we'll understand the dual-use of LAP in physics. To do so we'll go back to Feynman Diagrams (FD), where the 'Quantum-LAP' is fundamental in a way that cannot be evident in either classical 'big' physics or GR. FD are used by physicists to make precise calculations for probabilities of quantum process and events, of 'minimum-actions'. FD represent interactions of particles; and within those interactions, the least possible (events or) actions in nature, there are no distances nor accelerations, and no passing of any time. None of these three, including gravity, of course, just do-not exist on the lowest levels of quanta. Now for the punchline. Within the FD, particles are allowed to, must even, go both forward and backwards in time. Yes. Such physics smoothly leads to ideas like the Quantum Handshake, to a metaphysics where some events, some minimum descriptions of reality, become the very-image of particles, and a physical measurement of – time. How mind-opening do you find that? It's also the first time we meet time in its most profound sense. A-lot more on time in Chapter-6, where it'll be the right space and time to speak deeper on this (non-linear) 'feel' we humans have for events that are passing by. But now, let's review our understanding of what exactly those 'point-like-particles' are.

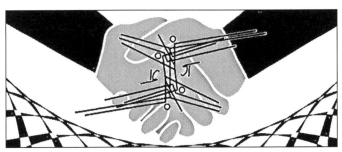

We have just learned of a physics' term and idea that is as fundamental to science as: "0+1+2=3". No less. The LAP is both an assumption and a fact, likewise the manifestation of energy as mass. Without this laziness of nature, assuring us some predictability of paths taken by stars and plants and particles of light, for example, we couldn't build any theory or model of physical nature. We also saw its dual-use. The 'Dark' in cosmology or 'Hidden' in QM are used for somewhat similar 'states' (the lack of knowledge), only in separate areas of physics, at least. Unlike these two terms, the LAP means two relating yet quite different things, and it can concurrently-dance in two far-away weddings. The other dance-floor the LAP dances on, besides the 'laziest-trajectories-rave', is in the realm of the quantum. In QM/QFT, the LAP means the most basic events physicists can possibly analyze, just like the event-of-change represented in the derivative of a moving vehicle.

As I've tried to show, and now I conclude – the example of the derivative indicates-much regarding the math of our mathematicians. It displays the at-its-heart quantization; the quantization of an event (or action) that is beyond the scopes of space or acceleration or passing of time; whether for changes in functions, or for minimum interactions (or facts). Beyond these least-events, deeper, scientists are just unable to (currently) explore. Maybe *Fractional Calculus* will allow us to drill-deeper-down. Probably, some entirely new math, perhaps delivered by a 'New Newton or Leibniz', is required. Here a question arises. Is it the nature of nature, the wave-particle-point-like-quanta, or is it just a limitation of our current mathematics? I promise an answer, at-least an educated guess, just a-bit later. For now, I'll focus on an answer to a much more important (for you and I) question. What on-Earth are we, actually? From this point of this read, things are becoming extremely metaphysical and quite speculative; however, non-detached from the so-far stated-facts.

Entanglement(ing)[1]

Maybe, probably, the 'punch' of the last couple of paragraphs wasn't felt enough. So, allow me another swing at it; before I'll try to portray how the point-like-stuff we are, 'entangles' into us and our dogs and our beloved flat-screens, small or big. We saw that everything is made of these things that can only be described as points-like, whether due to reality or the limitation of our math, or humanity. These points also possess several attributes, such as Charge and Mass, and another, called – a *Quantum Spin* (Paul Adrien Maurice Dirac was the one who concluded that it is needed, and therefore it must be real). These point-like-particles, the matter and force ones (that matter-particles 'exchange' in interactions), all 25 (they say) that exist, don't really roam free or able to be individually seen. Even in the LHC, where particles are found, they 'exist' for much-less than a second, before disappearing. It does not advance us much, logging such-short-lived particles; but, it makes our physicists so-very-happy, which is also quite important (yet to an extent). We've also realized three major points, regarding the point-like-elementary-stuff. Let's swiftly gloss over them, before we'll build some cool ideas on-top of these realizations.

---

[1] This has nothing to do with the album and/or movie: "ENTANGLEMENTING – An American Ending" (2019)

The first, is the understanding that everything in nature is constantly vibrating, and moves awfully fast. A common *Hydrogen* atom, for example, the most abundant element in the universe, is more than 99.99999999999% empty. This last statement is not true, and directly it relates to the fact that some call an electron a *Probabilities' Cloud*. This is where the WPD, the WF, and the beyond-human-time velocity of particles, collide and knot-up. That's why it is wrong to say that an atom is empty. It is not. It is just made of fast-moving stuff (faster than human time, one can say), that in our (average) experience seems as if it is all connected into a solid or liquid or gas. Matter seems full of stuff, and it does; but, it is no more than the fast movement of physical points; that, in our eyes (or time), seem as if they can be, and are, in two places at once, smeared as a cloud of potentials. More than anything, the WF, that haze of probabilities, defines the borders within which a particle might be observed. These borders are what define reality, above it all; and in a span of the minimum-human-time (around 0.01 second of a seeing experience), a particle will be found everywhere inside, all-over the space of the known information's borders. There are entire-worlds within the hazy clouds of particles. Imagine the cloud-of-interactions being everything inside the *Atmosphere of Earth*, and all particles' events are what happens in and on our home-world. Yes, even Earth is bordered, and we and the things we desire, it all exists within this quantum. Only when looked-at up-close, one can distinguish them.

Why Science and Psychedelics Go Hand-In-Hand | 97

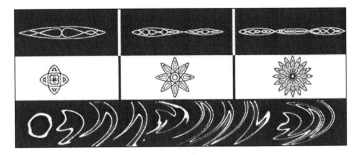

Now for the second and third points we've learned about those point-like-things. The second is that these never exist on-their-own, but as an (inter)action in a bordered space of-some-sort. And, remember, in these quantum-events there is no distance nor time. This directly leads to the conclusion that time is emergent, a result; or just a metric by which something else, some-other-thing of more fundamentality, is measurable. The third point on points, is that of Entanglement, of beyond space and time synchronicity of matter. Einstein and his (Jewish) colleagues, Rosen and Podolsky (together *ERP*), were certain in 1935 that they're about to destroy QM's theory. But, Mr. Bell suggested an experimental mean, statistically-heavy, to test the missile that ERP had launched at QM. He opened the door to sterilize ERP's argument, and for the establishment of a much-deeper, 'dark' and 'hidden', understanding of reality. It's hard, but that's a fact we must accept, for-now and until further-notice. Data travels at the speed of light, light carries it, indeed; but, there's something else that allows a faster-than-anything synchronicity. That, bluntly – it is us.

Here I'll connect the dots (sorry, the points), just like kids (and adults in 2021) in their drawing-books, and present what I understand are – us. About 99% of our body is made up of hydrogen, carbon, nitrogen, and oxygen. We also contain smaller amounts of the other elements that are essential for life. We are simple, basic even, quite fundamental. So, how do atoms, and atoms' point-like-particles, all these quantum events, end-up becoming us? Entanglement – is the best answer out there, as I, with my limited-mind, get it. There is simply no doubt that matter can and is really connected in means that our limited-minds find illogical. We are completely aware we only witness a fraction of physical reality, for example. I'll elaborate on this in the sub-chapter on – Seeing. What makes us think we 'see' all the rest, where there's clearly an evidence of entanglement? Now I'll briefly lay for you what metaphysics I assign to the facts, considering the history of the evolution of our understanding of reality, including us. The idea below will only begin to be developed here, and shall slowly evolve till this read's end. From this point, all points, everything we see, think-of, are, will be unfolded as – Entanglements of Data.

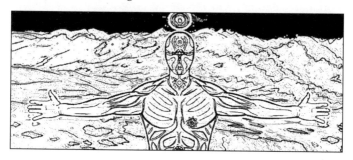

The glue that binds us-all as-one, as individuals and societies, just like Gluons are (bosons of the strong force) said to literally hold atoms' nuclei together, is Entanglement. This beyond any local cause & effect, this synchronicity on some 'another dimension' or Variable, is as fundamental for our (fine-tuned) existence as the Least Action Principle, or the hexagons being formed by dense-spheres. It is induced from quantum-facts and some scientific logics, and borrowed (by me) to higher structures. I have come to accept, through reason, that what is binding wave-particle-hazy-clouds together, into structures of many kinds of manifestations, are (some-kind-of) - **Elements of Entanglement**.

The evidences and the strict rules, which our formidable scientists find after lifetimes of experiments, must always be right, and everywhere. Maybe 'everywhere' does not really fit here, since it's becoming more imaginable than before that very-distant locations in the AetherFluid might exhibit not-the-same physics. No-longer this is an impossibility. But, even if there are only 'local-rules', in-case our cosmos is more heterogenic than even imagined, such a reality should be the result of some higher-level 'Software', a higher set of rules, for a reality of an emergence of 'local-realities'. Things might seem (relatively) different from different perspectives, but the rule-of-rules are unchanged. Quite like the thinking-of-thinking. In the only existence we're aware-of, no recursion is ad-infinitum.

As an economist, who was taught micro-and-macro (economics) for two degrees as if they stand-divided, I recognize the issues with unification. The subject of unification is problematic from two opposing sides. On the one hand, breaking reality down into few basic models is what our researchers of physics and economics, for example, do for a living. If you take away from them their models, they remain with nothing to play with. From the opposite side, a grand-unification of reality will always be a metaphysics. It is already becoming a widespread belief among physicists that the universe expands faster than the speed of light, placing a serious 'stop-sign' on where we'll ever be able to probe.

And neutrinos, those ghost-like-points, these are considered by many as more abundant than photons, and maybe even faster. So, who knows how those really work? That's why we'll always have to deduce rules, and even truth, from **Indirect Facts**, just like they do. It just must be allowed to everybody, and not only to those with the right colleagues.

Nevertheless, one thing, there is a fact, that is more than certain. Relatively speaking, on the highest of levels of physical rules; there is no time, only existing or not, interacting or not. Non-existing or Entangled – that is physics. And, while existing, during that event, there is no space nor time (like a Feynman Diagram) from someone (or something) else's, some higher, perception. Relativity is of a deeper meaning than previously alleged. There are 'Weaker' and 'Stronger' Relativities, just like -1 and 0 for 1 (either as real-numbers or concepts). The weaker is the local one, bounded by the speed of light (let's say it like that). But from the highest perspective, relative to the strongest relativity, space seems empty, like an atom, and entangled-matter seems synchronized, like atoms' sub-atomic particles.

Finally, just a quick sentence regarding the mathematics of our physicists, and their approach towards numbers and equations. Watching Brian Greene's Episode-5 of his "Your Daily Equation" (YouTube) series, is simply a must. I'm very surprised it is still on-the-air, and wasn't scrapped. In it, Greene, a 'physiceleb', tries to convince the viewers that 1 equals 0.9999999999999999999999999999999999... Yes, I shit you not, and, I think it's an accurate representation of physicists' (easily-bent) approach towards mathematics.

Chapter 5

The Psychedelic Episode

It's about time. We'll talk some drugs.

It was taken as an assumption that the reader, that is you, is a psychonaut. A psychonaut of either meditation or psychedelic drugs; or, a navigator who knows how to explore The Self by utilizing both tools. In this chapter, we'll start to investigate the relation among minds, psychedelics, physics, and consciousness. There's a section in the chapter after the next, which focuses on consciousness from a meditation and philosophical perspective. But here, finally, let's talk some drugs. Psychedelics, such as Psilocybin or 5-MeO-DMT, are unique tools for researchers who investigate the metaphysical and neural origins, the complete nature, of – consciousness. The most compelling theories of how psychedelics exert their effects on the human-mind, point at an increase in complexity, and overall, brain activities. This leads to highly synchronized and multifaceted 'neural-networking'. For the sake of unification and simplification, and generalization (might-be-even), we'll define this neural activity as (no-more-no-less than) – the processing of data. If you think something else goes-on in that brain of yours, please do let me know. I don't think any other activity happens in our human minds, brains, besides the processing of data, either of internal or external origin. Nothing else. We shall see in two chapters how this higher synchronization of data-processing is associated with entanglement, clearly; and how such a view might grow into a model, allowing the development of a tool for quantitation of consciousness. Undeniably, all roads lead (me) there, to – Levels of Levels of Levels…

Speaking of consciousness, I must jump-back to the Double Slits Experiment (DSE). Too-long I have kept unattended its most critical aspect, the most profound fact of reality it had revealed, the way I grasp it. There is some unintuitive weirdness that must be mentioned here, prior to us venturing into Seeing, and what happens in our minds, those data-processing-brains, when on-psychedelics. I'm sure you still hold in-mind, since you are paying attention, what the DSE is all about. I hope you also remember its first main insight. Fermion-Matter-Particles are also (merely) 'Hazy Clouds of Potentials', just like the photons of light (Boson-Force-Carriers) are. Now, two more facts regarding the DSE, such semi-insane-ones, that one can build an entire description of reality upon (as I'll do).

First is the following. If the experiment is done with some-camera placed near the slits, in an attempt to capture "Which-Way" (this experiment's name) the particle 'took' when it 'went' through the two-slits (between the light's source to the absorbing screen), the experiment's result changes. Suddenly, and as unintuitively as anything could ever be, the result on the screen suggests that the particle is not any-type-of-wave, or cloud, but rather it's much more like a tennis-ball. Yes, measuring the wave-like-clouds, their 'Potentials', forces them to become (point-likes) 'Actuals'. The actual-particle continues its flight towards the screen un-interfered, thus the end-result changes. Observing waves, logging data from information, creates actual point-likes form potential wave-likes. This fact, somehow, is serving two very-opposing metaphysics. Bohr's WF-Collapse, AKA the Copenhagen Interpretation (CI); and, the Many-Worlds-Interpretation, in-which there is no collapse, just branched-worlds where all happens. The second fact I wish to add of the DSE with single-point-like-waves, is that results are similar with some atoms, and even some molecules. This means that much 'bigger' matter, higher-above 'point-likes', is just a Potential prior to it being observed or logged, before it is Data. And, very recently, they started arguing over an **Entangled Tardigrade**, said to be the first 'entangled-animal'. It'll certainly reinforce the suggested metaphysics here, if it'll receive the stamp of – real.

From this point-on, points should really start to sync and emerge as a vibrant picture of reality, way-above the entangled-emergence of you and I and all matters. This is so, since we, humans, will be at the center of the analysis.

It's time 'to-go-meta' beyond those point-like-particles, or ripples and waves in the fabric of spacetime (which we'll be back discussing metaphysically in the coming chapter). Our nature will therefore be the prism through which we'll break down and then reassemble nature's nature, as it can be derived from several facts and experiences. The physicists we adore should (and do) receive all the respect in the world, but, it's time they'll think just a pinch-more of humans. 'Science' is worshiped by scientists no less than the Sun was by the Pagans. But what is 'science', if not theories, models, and munged-data? And what is it historically, if not a time-limited understanding? Finally, how can it be that science (& tech, as an extension) have become a 'public-enemy[1], in eyes-of-many, during these COVID-19's days? The one answer I tell myself relates to my recent imploration. Scientists do not consider, nor respect, humanity and societies enough. Money, fame, glory – are yet-again the golden-cows of our leaders.

---

1   I don't mean the New York hip hop band that Chuck-D and Flavor-Flav formed in 1985.

Seeing

"How can one tell of what one saw while tripping, without first defining what is Seeing?" I'm sure I've heard this question somewhere. And even if not, and I'm just hallucinating, falsely-remembering, it's a question worth asking. I've already taken a risk and told you we are average-seeing-creatures, (by-design) able to process data from some-certain level upwards. Else we would've been some-entirely different creatures. Here I'll take another quantum-leap, (or better) a quantum-risk, and clarify some aspects of – Seeing. We just can't move-on without it. You're seeing these words because light shines on the page, or, from an eBook. Then, data is transmitted, via photons, to your eyes. There's a chance that your eye and the page meet-half-way, beyond time, like a Quantum Handshake. Later we'll toy with that too. We receive eye-data in the following manner. There's a source of an electromagnetic radiation (light), an investigated object (page), and a detector (eye). And, our eyes and brains see only a tiny-part (~0.036%) of the electromagnetic spectrum.

When it comes to (not only) seeing, we are quite limited in our ability. Radio-waves and Microwaves, Infrared and Ultraviolet light, X-rays and Gamma-rays. Those all are light, all of them are photons, yet not all photons are visible to us. We will never be able to see atoms, however closely we look with a magnifying glass. They are smaller than the wavelength of 'visible' light, and thus cannot be seen under any ordinary microscope. Light is the radiation given out by the Earth's nearest star, our 'Mother Sun'; and humans have evolved to register only some very-particular photons. A quick analogy with sound. Imagine a piano keyboard with a rainbow (red-orange-yellow-green-blue-indigo-violet) painted somewhere in the middle(ish), between the infrared and ultraviolet. In the case of sound, you hear the whole range of octaves; but, the light we see is only of the rainbow that's painted on the 'electromusicpiano'. As we go over the rainbow, from red to blue for example, the wavelength halves, i.e., blue's wavelength is red's half. Surely, "what a wavelength is?", you ask. That's the distance between two near crests of a snapshotted (from the side) wave, I answer. The more a wave is dense, its wavelength is shorter and more condensed. A gamma-rays-laser will be the strongest, since it'll shoot a stream of very-dense-energy-packed photons. But, that's not the 'seeing' I was really meaning or planning to talk-of. The Seeing I did wish to explore, is completely entangled with Time, and, undoubtedly I believe, with the fact of the reality of – Entanglement.

I will start by uttering an obviosity. A 7-year old kid, even though having by-then a fully developed set-of-eyes, will see differently from an old-geezer at 75. This is 100% due to entanglement, or, at-minimum – Elements of Entanglement. Don't you go assuming that even the top-tier brain-scientists alive, are definite what *Memory* is, what is it 'made' of and how it is 'stored'. None of the above is certainly known. So, I'll make-up my own (god-damn) mind, just like I always do, building on the agreed upon evidences. There's a straight line between seeing and time; and, between time and the density of space (the AetherFluid, that is). Much on this in Chapter-8. Now, some more on the former. Just like they do in Hollywood, the movies I mean, existence in (by definition, and design) snapshotted from a human perspective. This statement is correct twice, in my mind's eye.

The first reason is that 75-year old humans, when witnessing matters and things, i.e., absorbing photons, also see in them their personal experiences and the social-norms they are subject-to. That's entanglement. The Information and additional Data, associated with photons' data, is Entanglement. That's unity, beyond space or time or the speed of light. Simply put, all is one, from the perspective of, let's say, the biggest blackhole out-there. It is all a matter of synchronicity and entanglement's level. Just like a movie, every frame is connected to, inheriting from, the previous, and clearly leading to the following. Same goes for scenes, and for (mostly lousy) sequels or prequels. In the (surprisingly clear) words of Bohm: "Each moment of time is a projection from the total implicate (or enfolded) order." Just like quanta of fields, that can only be seen as point-like-particles when observed-at, interacted-with, as such; so does the case with snapshots, quantization of time, and time itself as-a-concept. Alone, a frame or image or moment, is meaningless.

The second reason the statement from the paragraph before the previous holds, will also be illustrated with the help of Hollywood. Just as with their movies, even the 'copy-paste' ones, data is delivered into our eyes and mind in (entangled) frames, that only together their meaning is derived. Every frame is not only in-sync with that which came before and all that'll follow; each holds an inter-entangled (photons') data, and their separation is merely apparent, a result of ignorance, no-less. I've mentioned a while back that time slows down (relatively speaking, always) for a moving clock or one that's at an area with a stronger 'pull of gravity' (though nothing is pulling anything). The subject of gravity will receive some treatment soon, but a quick note on movement now. How did Einstein (logically) prove that time is slowing-down for moving matter, compared to a stationary one? For Special Relativity's sake, he imagined two parallel mirrors positioned one in front of the other in a moving train, or Uber, or Lyft, or whatever. He concluded that a photon, trapped between the two mirrors, endlessly flying back and forth, will hit both mirrors fewer times than a photon in the-same set-up but in a non-moving vehicle. Yes, events are at the heart of everything – of space and time and experiences of matters.

Indeed, the processing of data, of the happening of events (interactions between a photon and mirrors or your mind's eyes, as examples), is not only time, but it's existence itself. Our 'feeling' of time is in direct correlations with the 'number' of photons we absorb. Time is absolutely emergent, i.e., it's non-fundamental and subject to perspectives. And, non-fundamental or perspective-based 'facts' – these are, more-or-less, of the same value as of our politicians' campaigns and 'honest-promises'. It is fine to be-aware-of, to know-of, yet shouldn't be taken entirely-serious. Seeing is, now I conclude, the receiving of entangled photons, inter-entangled into an image and in-complete-sync with past and future, and with everything ever experienced by the observer, in-and-out-of-context. This creates meaningful-frames and other pieces-of-data that are processed in our minds. And this processing, like the bouncing of a photon between mirrors, is – Time. A brief sentence about – 'Observing'. In physics, for physicists, *Measurement* is a tremendously charged concept. It "triggers" them, in the Gen-Z's language. Why is that so? The answer is, yet again, the same. They just have-no-idea what happens during the process of a (QM) measurement, in this data creation. They don't have a physics for – Actuals of Potentials. In my opinion, and from my (psychedelic) experiences, it is all for and of – Consciousness.

Psilocybin

This one is a personal favorite of mine. Been making-sure to consume this substance yearly, twice occasionally, for more than a decade now. I find it healthy, greatly. For body and mind and soul, all-at-once, as one. This is not a recommendation, nor an advice. I simply wish to tell you, provide an in-context 'trip-report', of my experiences with this fine psychedelic. The perfect one, in my own mind. This natural-chemical, that some believe had an impact on humans' evolution, can be found in many of what is commonly known as – Magic Mushrooms. Usually, these powerful cuties grow on animals' manure. Among the 'Shrooms', **Golden-Tops** (*Psilocybe Cubensis*) are my most beloved. Psilocybin is the most common cultivated psychedelic, and it can be found in the wild in many places, worldwide. In Australia, mainly in the tropical part of the continent, but surely not only; I have spent several months 'hunting' them, studying psilocybin's effects, before it was added to my (yearly) diet. One thing's for sure, both seeing, and data processing, are becoming fundamentally different once psilocybin is consumed.

Images of the human-brain while on psilocybin are easily found online, and the firsts can be dated to the 1980s. Even a non-brain-scientist will get the picture (or image). These are produced using a functional MRI (*fMRI*), which detects changes in blood flows of (or in) brain activity, showing degrees of mind's connectivity and synchronicity. Everything, any measure and metric quantifying a facet of brain-activity, is growing (exponentially) with the consumption of psilocybin. But that is 'only' the (brain and mind's) physicality of the effect of this molecule. The complete human experience tells more of this substance, and of ourselves. When it's combined with other facts of physics, and several ancient (and advanced) notions and philosophies of consciousness, a complete picture of reality forms. Therefore, I'm pretty confident that psychedelic-substances and psychedelic-experiences should eventually, sooner-than-later, find their way into physics' labs. Let's examine that.

The growth of all brain-activities while on psilocybin, is in complete correlation with data-processing. Just like the heart is pumping blood, the brain processes data, and that process of data-processing grows with psilocybin. This is a fact. Now let us build some metaphysics upon that. There is an aspect of psilocybin-tripping I wish to share and analyze. That is the experience of seeing while-under. I have mentioned it before, that everything becomes wavy for me (an hour or so) after I consume it; and I don't accept the explanation of hallucinations. Every idea on drugs formed by the Baby-Boomers should be flipped on its head, no-less. How can it be that the real nature of nature, the constant fluctuations and waviness of matter(s), is surfacing in-correlation with data-processing (which is positively correlated with psilocybin consumption)? Speaking in Boltzmann's image: Brains, which clearly 'work-harder' when injected with a stimulating psychedelic like psilocybin, able to perceive beyond the on-average mundane human perception. Yes, (to me) it's clear that psychedelics are not hallucinogens. The definite opposite seems the truth. Psychedelics reveal deeper aspects of reality, beyond the limited senses-perception that fitted us. That is – us. I can't say I understand the mechanics, but I can say I was aware of the fact that I literally see more. I can (almost) say I was just processing more photons. Only concepts of 'God' can be imagined as an 'Ultimate Perceiver', observing and processing reality in its entirety. Later more on that, about some physics for – God(s).

But just a paragraph more on what is the human-feeling of such beyond-mundane seeing. I can only speak of my personal one, the only feeling I'm aware of; but, all I read and every psychonaut I discuss this with – strengthens my analysis. Back in 2008, before psychedelics or psilocybin started tiptoeing into the mainstream, I had a couple of pretty profound psychedelic experiences. In the first of this series, I've eaten a few big and crispy dried golden-tops; and within less than an hour – I was a tree. "Was a tree?!?", you attack, pseudo-ask, unhappy with such a ridiculous statement. Please, allow me to explain. With this 'higher-resolution' of perceiving, the additional external data flowing into my brain via photons (or the enhanced processing capabilities), both in quantity and (as a result) in time, came a feeling of connectiveness and wholeness with what was perceived, with where my attention was laying. While I was raving at the singular "Freakreation" festival on Boonoo-Boonoo Island (Australia), way-before I heard anything about the quantum, I knew I felt that shrooms made me, somehow, 'see-more'. I was staring at a huge wild tree in awe, completely aware that suddenly I see more leaves at once. My mind then was not thinking of minds that are processing data or models of consciousness (beyond my then 'religion' – Advaita-Vedanta). The more I saw of that tree, the more attention I gave to the photons that carried its data, the more I became it and only it. That's also part of the beauty of psychedelics, of levels of consciousness' changes. Once a drug drags one from everyday experiences, from us, we're naturally 'pulled' outside our own identities, and into some higher, wider, human existences. Nothing does this more than – **5-MeO-DMT**.

5-MeO-DMT

This drug, not much after that fine experience with psilocybin, has provided me with the most mind-bending experience I've ever undergone. Most-probably, as mind-opening as an experience can ever get. Plain words won't describe well.

Still, it is my job here to try. This psychedelic comes from the venom of the *Colorado River Toad*, native to the Sonoran Desert (on the southwest US-Mexico border). It's an extremely potent psychedelic, about five times more powerful (or impactful) than its drug-cousin – **DMT** (*dimethyltryptamine*). This liquid is extracted by milking the toad's toxic venom glands, and then dehydrating it into a crumbly dry paste. Shamans throughout Mexico and the southwestern US have been harvesting and smoking this substance for decades, and many thousands of people throughout the West are now seeking-out this effective substance, and its experience. Yes, even this 'radical' psychedelic has become trendy, and it is certainly not a party-drug.

I've tasted a synthetic version of this molecule, administered to me by an Australian-Jew who's also a psychologist by profession. The session occurred during the final night of a festival called – **Maitreya**. Unlike psilocybin, which tripping-on might last a couple of hours[1] (still less than *LSD*) and is felt in the body and brain for several months after; the 5-MeO-DMT experience is a less-than 30 minutes kind-of-trip, and what truly remains 'in the brain', is an image of a filled-with-light realm. Furthermore, the neural signature of this drug shows an increase in *Delta and Theta Waves*, normally present during sleep, and, even more particularly, while dreaming. That is why it is quite-commonly called an out-of-body-experience, one beyond any other psychedelic trip. 5-MeO-DMT appears to be associated with anti-addictiveness (as it was so for me);

---

1 This can be managed with the addition of citrus. The more lemon you add, the shorter and stronger the trip is.

and, a sustained enhancement of satisfaction with life, with being alive; and such other mindfulness-related capacities, including improvements in depression and anxiety. Despite these advantages over other drugs, 5-MeO-DMT remains scientifically neglected and underexplored, and no brain-imaging studies, or vast clinical trials, have to date been carried out with this compound. It is time that is changed; and better sooner, much-much sooner, rather than later.

Now to my own experience with it, which is in-line with most trip-reports by the most famous-of-trippers. Flatness (non-dimensionality), inseparability, and a complete lack of 'normal' processing of data, and consequently a zero-feeling of time – these all describe the weird experience that 5-MeO-DMT had provided me with. That's the uniqueness of this drug, the fact that the experiencing-of-it means to go beyond any-type-of seeing, or other aspects of 'mundane-being'. A second after I've inhaled the thick smoke from my glass pipe, and as an answer to a friend who asked to know of this psychedelic's flavor, I replied: Orange. I didn't mean the fruit; but the color. I can only explain this by assuming that 5-MeO-DMT is either unifying the five-senses, or it is shutting them down, allowing another to dominate. Another second had passed, and I went away. Or deep, maybe. My body remained in the place I was at; but me, my-self, the observer and perceiver of events and entanglements, which are I, was somewhere or something else. Like a dream, where you're just another character in a beyond-space-and-time scene. If you'll survey many of the trip-reports found online, it is evident that people are 'seeing', and being, geometric shapes of many types and sorts. Me, I was a honeycomb. All I saw (with eyes-wide-shut), were endless dense hexagonal shapes, where each several neighboring cells were colored differently, by all the rainbow's colors and shades. I've spent four years contemplating that experience, trying to figure out (just a-bit of) what the-hell was going-on there.

It all came down to the novel I published – "That's Not Thinking" (2018). In my book (and life), when I was slowly coming back to my senses, truly-literally, while I was still in an existence where 'Boaz' had no meaning, still with no identity nor a memory; I felt my cleanest of thoughts, that, felt like a message. I even started speaking it out-loud, with no recollection that so I have done. It took me around 30 minutes more to normally, I'd better call it – humanly, sense my own existence again. I completely understand that it all sounds like one-big-hallucination. But, how can a human experience be neglected, if we are fully aware to the fact that human-beings are only aware of averages, and just a thin slice of the electromagnetic spectrum? And not only that, and by now it's completely documented. Psychedelics increase mind-activity, and allow us-humans to, momentarily, 'step outside of themselves', out of our matter and social entanglements, if only for just a short-while. When doing so, we get detached from our mental illnesses and many other mind-sicknesses. When coming back from this detachment, new proportions are formed. Relativity is everything, mental-including, and it's important to harvest those in-between perceptions for self-reflecting on ourselves and our life's facts, differently – for a change.

What else can I deduce from my fine 5-MeO-DMT experience, if combined with some of the physics' facts surveyed here? The first which comes in-mind, is the reported-by-all flatness. You see, but the only 'thing' we can imagine in nature that's completely flat, is light. Photons, you know. "Why so?", maybe you're not sure. That's because those light's photons have no mass. "And that's why?" That is because they move in the speed of light, I answer. A circular argument, isn't it so? Both the law of *Energy & Mass Conservation*, and Relativity-Theory, are like that – quite-very-much-circular. Light is the reference frame of our (photons' dependent) existence. Photons are **Nature's Denominators**, some kind of index-definers of Mass and Time and all that is determined within and by Space and Energy. But I am getting way-ahead of myself. Just a few more maneuvers are still needed to get us there. As promised, bit-by-bit, point-by-point, an image-of-all will be molded.

The other facts of physics I wish to integrate with my own experiences with drugs, both 5-MeO-DMT and psilocybin, is, quite obviously, of – time. When one 'goes' to some 'place of no time', at the peak of a 5-MeO-DMT trip, it seems like a second. I couldn't believe that it has been more than half an hour from the time I answered my friend's question till I spoke of how I'm clearly seeing, for the first time, my own ego. How could I have really tasted and know of it, of me, without knowing its opposite? I'm only human, still. But not only a place of zero-time makes my case here. A psilocybin trip is also a great example of the physics we live-in. The heavier a trip is, the more shrooms and lemon one is mixing and consuming (resulting in a denser mind-activity and in-depth seeing), the longer the experience will feel from a human perspective. It seems as a fact – more data-processing is correlated with the feeling of more time-passed, and no processing of data seems like an out-of-time existence. We will soon see how data, light, and time, are one.

Necessity

The question of psychedelics' necessity arises. How necessary are nature's drugs to the ordinary person, the scientist or the psychonaut (who should be quite scientific as well)? In the most technical of terms, how can we imagine the very-optimal role of psychedelics, and other drugs, for personal and social benefits? Well, we already do. We just chose the wrong drugs. You know why, mortgages (and cows) and such. And all those low-interest mortgages are being paid with the 'good' drugs only, and not the ones coming from toads or cacti or, literally, (fungi on) pieces of shit. So, this question in a non-one. Everyone agrees that individuals reach mental states that can only be aided with the help of some molecules, and that is fine, making perfect sense. But, playing the FDA's game of which ones, is a game of money and political-power. Which leads me directly to the next point.

Every psychiatric will say that sometimes some-drug is needed to pull patients from their own-selves, from attachments to their repeating thoughts. But what of our modern, tech-attached, quite sick, societies of – **Social Networks' Relativity**? The top-cause for contemporary youth's depressions, is the self-defining through online networks. There, just too-many 'facts' and almost all images, are fake-news. There is a dual falseness there. The photons carry lies, the images are falsified; and, the entanglements they produce and arouse are non-real, to say the least. A modern-human is entirely engulfed is a reality of abstractions that mean nothing, or very little. But, when there's nothing to compare to, as the (on-steroids) relative nature we're stuck-in constantly demands; all proportions are lost regarding what's real or not, and what perspectives are of value and whose are not. As a technicality, higher (psilocybin) and/or altered (5-MeO-DMT) consciousness' states, provide real-relativity – the kind that I find of great value for humans. In our materialistic and mental existence of *Social Relativity*, with never-before-seen amounts of data-points and not-enough-information, real proportions are valuable. Priceless, even.

Which brings us back to where we've started, to our great scientists and their science. So, why do science and psychedelics go hand-in-hand, as I perceive it? As almost always, answers can be delivered from both sides, like a handshake. Science and scientists need us, twice. First, psychonauts are needed as guides, aiding those who wish to experiment upon themselves. I hope they will explore their true-selves as they would've done with any other research, by consulting those who've been there before, reading our words, and learning from our mistakes (most of all). Secondly, scientists, physicists including, should allow the human-experience a central place in labs and experiments. Have we learned nothing from the DSE, and its clear indication that nature is even lazier than presumed; remaining as information-only, as potentials' waves, as-long-as no data is logged by some observer (via an interaction of a measurement of-some-sort), when it becomes an actual? But physics is still contemplating the nature of measurement. Yes, I know, it's the 2020s; and no, I shit-you-not. They're not sure what's a measurement; and confuse between the measuring apparatus and the quanta it is made-of, undecide which of the theories they believe-in needs to be applied here/there. Why not just take a step back, inside, and first understand ourselves, begin by building a fitting model of humans' interaction with the universe? This is true for the personal as the collective, as we will see towards the end.

And we? We need science to keep us, and the Colorado River Toad, safe. Allow me to explain. We live in an era of grave misuse. There's no doubt about it. From coffee to psychedelics, through pharmaceuticals and even cannabis. Just an age-of-misusing. Sure, I include my own misuses; and surely as-well, I'm fully aware that too-many humans are currently alive for this planet to handle. It is known. But, it doesn't mean that quantifiable and scientific approaches, as in science-labs, won't aid with regulating all the substances that we, psychonauts, cherish. I'll elaborate. Regulation is also men's friend. You know where there's none? In the jungle. And you know who rules there, and how exactly, right? So, regulation is simply inevitable, and it requires laboratories, within them experiments happen, allowing for specific molecules and doses to be formed and optimized, and for government-asses to get completely covered (which is also important). That is not going to change any-time-soon, and that is not the problem. And, in a sense, we are not helping.

Sure, I understand the value of consuming natural-only psychedelics, in their natural setting. But, it is unsustainable, whichever way you look at it. It has nothing to do with the pharma companies. New, psychedelic ones, are forming and will-be formed. But, we psychonauts must take it more seriously. Speaking in language of *Climate Change Science*: "We fucked it up". So, let's try not-to this time around, ok? Take for example the Colorado River Toad, that its numbers are negatively correlated with the weekly google searches of the term – "DMT". Come-on, we all know exactly how it is going to end, don't we? And that's not all, and here comes the main preaching of this effort here. The marriage of psychedelics with science (where psychedelics is clearly the bride) must result with a much-more scientific approach to self-experimentation with drugs; and a more vigorous philosophy of consciousness than those found at the heart of many New-Age concepts. Told you already, I find the New-Age ('philosophies') as baffling, and sometimes even purposely misleading. A well-established (and a multidisciplinary) **Science of The Self**, one that's incorporating more established and traditionally accepted, and even regulated, methods of self-exploration, will lower the fakeness in (modern) 'Spirituality'. If it's not clear, I'll explicitly state it here: We, psychonauts, must become much-more scientific; more than science, scientists, must become much-more psychedelic. That's my thinking.

CHAPTER 6

The Matter That Matters

It's about time. Things become.
Meta yet real. Here.

"What is matter?" was one of the more important topics for the Greek philosophers. With pure logics, and that alone, Democritus, somewhere around 400 BC, imagined the Atom (which means – "Uncuttable"). If you think, if you believe, that in 2021 we have a clear answer to the above question, again, you're grimly mistaken. There are many reasons why fundamental physics did not evolve for more than 40 years now. At least one pops in (my) mind. If there was a time-machine for rent, and you traveled back to 500 BC and consulted **Anaxagoras** of Clazomenae, he would probably tell you that even your future physicists still did not understand what is – Space. Anaxagoras, besides being first to suggest that the moon reflects light from the sun, presented a (most-relevant) theory of **Everything-in-Everything**. In his mind's eye, all matters whatsoever are not alienated to space, and consciousness is the motive-cause of existence. All is simultaneously in and of space, very similar to point-like-particles and waves; and, within space, events are a result of energetic-consciousness. Moreover, if you'd tell Anaxagoras that point-like-particles are considered in the future as the fundamentals of nature, he'd respond back: "But what if the world is not composed of material particles? What if there are no indivisible entities and no bedrock at any depth of the decomposition of reality? What if the universe is not atomic, but gunky, sticky, messy, with parts of parts, of parts, all the way down (or up)?" This is a concept that has been investigated little in modern physics, but Shlomo Barak did just that. Among the many innovative ideas in contemporary physics, Shlomo's are the most compelling, and certainly were most-encouraging for this analysis. One can say, that **Barak's Space** and **Rovelli's Events** are the physics (and physicists) I most follow.

Unification has always been the goal of brave physicists, and scientists in general. Unification is the generalization of all generalizations. Shlomo Barak believes he was able to achieve this holy-grail, and combine the 'Small' with the 'Big'. In his model, another avant-garde Theory of Everything (ToE), an electron and a blackhole are, more-or-less, the same phenomena just on a different scale – differ in ratio of mass to space. Same math, same equation, describes both of their physicality. As told, I came across Shlomo's work by accident, while starting to write this book. He has concluded that all of physics, all phenomena (and especially that of *Electric Charge*), can be explained with a simple new *Space-Geometry*, such one that allows dense-granularity, if combined with another new theoretical mechanism of 'getting-excited-photons'. It mainly relies on some-assumed, additional, never before seen, wave in/of space. All that exists, all that fundamentally there-is, can be described in a model of purely space-deformations. In his theory, that he called – **GeometroDynamic Model** (GDM), time is not fundamental, at all (as everybody by now should well know); rather, *Distance* and *Velocity* are. But, everybody also knows that velocity equals distance divided by time; so, for him time is merely a plug-number, which I totally agree with. "Time by itself has no meaning"; he gives time a sentence and wastes no-time on struggling with time. The best way to avoid an attack on something you had said, is by avoiding all that is puzzling the rest of science, and not saying it. The fact that bothers me the most about his attitude towards time, is that I think he is right.

Shlomo, as I've come to realize, would-not endorse these words of mine. Most of it, I think. You probably wish to know – "Why so?" That's because, like too-many physicists, he is ignoring language, and humans. Only 'science' and papers are of importance; and any statement that is not 100% spot-on, dry and sterile, and backed by a measurement, shouldn't be written. He's also of those believing there's no such thing as consciousness, that it is all "just an illusion", like most Baby-Boomers grew to believe. He declares, with great confidence, that he has fulfilled Einstein's vision and composed a theory completely compatible with Albert's spacetime-fluid. Yet, he provides zero-reasonings for where exactly energy comes from, or is, or to the unintuitive results of the (Which-Way) DSEs. None provided. He has recently (2021), and without a word on what is a measurement, started claiming he has provided a physical model for entanglement. He has not; but, just like them all, he wrote in one of his (2017) papers: "We do not know what time is, we only know what motion is." Indeed, more-of-the-same on inertia and/or movement. But, it will be misleading to present it as if he did not innovate, and I don't mean his enormous contributions to the Israeli defense (and attack) systems. In his vision and mathematics of a cellular-spacetime, many should find several groundbreaking and advancing ideas for known problems in physics. I sincerely wish him all the recognition in the world.

Bohm, unlike Barak (but in-line with Barak's own inspiration – Anaxagoras), realized that any unification must embrace the human experience, including some 'marginalities', such as art, inspiration, and, of course, consciousness. Sweeping time and consciousness under-the-rug, is no-way to go-about. Don't you listen to any technician who does just that. We have had quite-enough of inhuman-science. Too many billions have already been spent on such immoralities. It is time to unite, provide the supporting physics for our unity, and not to further divide. In the following brief sub-chapters, I'll attempt the act of merging of propositions. I want to try and speak in language and terms and ideas that Anaxagoras, Bohm, and Leibniz, might semi-agree with. It will start with a re-overview of our knowledge of spacetime-matter, of the AetherFluid; continue with a description of what happens within that matter, that matters; and finalize with an analogy for particles. Let's continue our journey.

An AetherFluid

We should go a few chapters back. Only this time I feel more comfortable to roam deeply and speak-philosophically, and quite freely. So, remember what they believe in? Some 14 billion 'Human-Earth-Years' ago, give or take, in a time where simply time-was-not, an Inflation event occurred, and I don't mean your rent, either you pay it or receive. Not that inflation. All that we now see and are, was condensed, without Relativity (I'm relatively confident), to a pulp the size of a volleyball. Not only what we observe was in that mush. Spacetime's foamy-fluid-matter was also crowding in-there, somehow. If considering the entanglement's fact, it is almost backed-by-physics to declare that we are one, since we were one, once. General Relativity (GR) was the last real paradigm-shift in cosmology. I do not consider Inflation because it is constantly changing. A pretty unsolidified theory that one is. It was Einstein who gave the matter of spacetime, the AetherFluid, attention, and was able to convince the world of some unity of space-time-gravity, supposedly. Currently, it is not so simple to determine which is more blundered, Particle Physics or Cosmology. Scientists who collide subatomic matter or those who look and listen beyond the skies. Since our leading cosmologists truly have no idea regarding 95% of what's on their plate, they justly admit, I crown them as – most blundered ever. Dark Matter (DM) and Dark Energy (DE) are beyond-blunders. These two are (no-less-than) bugs in the system and method of science. I had no idea to what extent this is so, before I met Shlomo Barak and was exposed to his work and vision. There's no doubt, the direction he took is the one that physicists (of 'small' and 'big' and all that's in-between) must now follow.

Einstein's GR gave the AetherFluid a fundamental role in our (physical) existence. Indeed, since 1915 and as of 2021, we are just (almost) 8 billion lost souls swimming in a fish-cosmos. There's no vacuum to-be-found, nor true-emptiness, nor real-nothingness. This was just before, 'Pre-Inflation'. We're always, everywhere, in-something. In Albert's vision, any fish or mammal or rock that swims in the AetherFluid, just like a whale or a shark or a whale-shark in an aquarium or ocean, deforms the AetherFluid's shape itself. This deformation is the cause of gravity; and gravity is manifested as geodesics, as routes, of planets and stars and even of some elementary-particles. There's no way to deny the (local) success of GR, but GR's general-accuracy is in question. It is not the same case as with the Standard Model, with Quantum Field Theory (QFT), where the lack of precision per-prediction is probably a case of Dark Variables – a lack of knowledge regarding what the theory and model should be aware-of. GR seems to suffer from a basic misconception, at least one that is the result of believing in a few inaccurate assumptions.

The need to invent and add a couple of old and new math-terms to the equations of GR, is an evidence of the fact that GR is not a generalization, but only an approximation, and quite a local one that is. Einstein's equations didn't account for the phenomena of galaxies' structures (DM) and the accelerated expansion rate of the AetherFluid (DE). Simply 95% of cosmic events, facts, are not in the scope of GR, thus it is probably not the most fitting of theories. Sure, it was great for describing the path taken around the Sun by Mercury. But come-on, isn't it time to rethink gravity? Can we imagine some of the traits that a fitter AetherFluid should possess, in a manner that it may unify a blundered area of physics while solving some general issues as well? One is – Condensation(ability).

Condensability was certainly the direction taken by Barak when he reimagined spacetime in his – **Barak Geometry of Deformed Spaces**. Conceptually, little-minor-me completely agrees. That was the energy behind our meeting, I was looking for such theories. Was surprised to find one a 20-minutes' drive from where I live. What Shlomo did was to start with a quantized space, his basic assumption, and atop these quanta, and with two additional types of waves, to build a model describing elementary-particles and their properties (plus some new-ones that no-one was able to even think of). But there's one aspect of his theory, that minor-little-me doesn't agree with. "The Sun and other masses contract the space around them," Shlomo finds and Einstein would've agreed. Me? I'd say that the Sun and other masses are colliding-waves of AetherFluid's streams.

Another property that a fitter-to-data AetherFluid should possess, besides being able to get heterogeneously-dense (which'll explain *Galactic Dynamics* with no need for any DM gluing stars or planets together) is – Fundamentality. That was how I became familiar with Barak, by venturing for theories where space is spawning, and is, everything. That is including – Consciousness; and vice-versa, in-a-sense. I hope we are in some-sort-of agreement regarding observations. It has been shown that, yet not how, measurements produce Actuals (data of events) from Potentials (information in waves); which leads to a realization that observations are as fundamental as the observables themselves. It's quite similar to the (number) 1. It gets its actuality from the 0 and the -1, exists in-between, relatively to them both. The accepted results produced in the Which-Way version of the DSE are in very strong accord with the above argument. They also exhibit the fact that observations are just interactions, (almost) by definition. In Baraks's GeometroDynamic Model (GDM), there is a unification between the electromagnetic wave of light's photons and space itself, resulting with two types of photons, corresponding to couple-particles of anti-matter and non-anti. And finally, in the GDM photons create everything. Indeed, in Shlomo's theory and its implication, all matter is made of light. I know more than a few 5-MeO-DMT's psychonauts, and many New-Age(rs), who'd surely agree with that.

Waves in/of Space

Waves of the AetherFluid, energetically moving and taking shapes, just like, well, the waves caused by an overweight dude bomb-jumping into a pool, are the focus of this section. Such were mentioned much before, the Gravitational Waves (GW) detected in 2015, said to be the result of two spiraling-into-merger blackholes in a far-away-galaxy. Told you, I have zero problems with indirect 'facts' and stories, with inferring what has happened during by (actually) observing only the very-end-results. But the campaigning and marketing of such a specific story is unacceptable, from a scientific point of view. It's too much of a storytelling. Having said that (and you should know exactly what'll come now), I'm going to do quite a similar maneuver. The word "wave" was used in this text pretty heavily. There's simply no-way of avoiding waves when dealing with quantum models and the fabric of spacetime. Waves are manifested in nature on the largest scales possible, and are also the nature, at least the behavior, of the tiniest pieces of matter our scientists were able to play-around with. If you think of it, just have a look around and in the mirror, it's quite evident that our existence is spherical. This is quite a conveniency. There is the subject in physics, of a (very) *Fine-Tuned Universe*, as our cosmos so-seems. It's kind-of straightforward, I guess. If one would ask me (though non-does), I'd say that the Mother-of-All-Fine-Tunings, the most-fundamental aspect of physical existence, is – *Sphericity*.

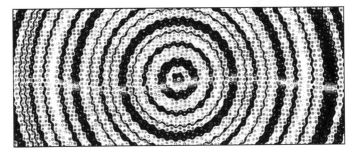

Waves are just a derivative of a spherical-existence. If you (Jehovah, Brahman, Allah or Buddha, or any other of Gods' names you're most familiarizing yourself with) wished to architect a wavy cosmos; a spherical-existence is a must, and vice-versa. It is not even fine-tuning, it's more like fine-architecting. This is since "tuning" refers to some variables, *Physical Constants of Nature* they're called, while nature's sphericity is analogically closer to the initial design of a structure, predating any tuning of variables. When architecting a universe, so I've been told, one starts by defining the variables the desired universe shall include, for the later-fine-tuning. The resemblance to the design of a *Game Economy*, or a new nation's economy, is clear. When I consult a startup in the free-to-play-or-earn industry, I always commence by the simplest design, with the very minimum variables needed for a sustainable virtual (or digital) economy of some casual (or crypto) game (or product). For example, the 'daily bonus' of some virtual coins in a casual-casino-app, is a pre-architected variable, just like sphericity, while its specific value is a free-variable that can change or be differentiated for dissimilar types-of-users. By the way, Shlomo Barak was able to architect a universe with the least variables possible. There are no-less than 26 constants in physics. Some are for several particles' masses. Barak was able to do more with 4, and the most accepted of them all: G for Gravity; C for photons' velocity; h for Max Planck's minimum energy; and, last but not least, α for the *Fine Structure Constant*.

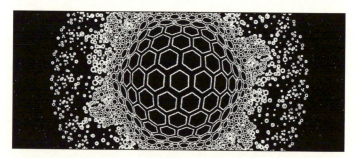

The sphere-shaped-cosmos we humans dwell-in, according to my imagination and lucid-dreaming, allows for the most important feature of physics, which itself supports some other fundamental aspect of our physical existence, the way I understand it. This vision that I have is also in-line with the current state of our 'mathematics-of-events'. Only this structure, such Universal-Geometry, establishes an existence of Events' Points. That's the reason why a bubble (not that of stocks and/or 'crypto-currencies' or any other real-illusion) is spherical – such can contain the most volume with the smallest surface. Laziness all around, again. But, when many bubbles collide with each other and get crowded, as in foams, the exact same phenomena that are found all-over nature, occurs. Some marvelous geometric shapes are formed. Structures that seem spherical only on the outside, but built of completely straight-lines, averages of spherical shapes, from the inside. The hexagon-honeycomb is an example. Only in a spherical existence there are waves-of-spheres that create 'points' of all sizes, from elementary particles to stars and planets, when colliding. Even the evident-from-up-close *Flatness of the Milky Way*, can be modeled as a quantized-point of a collision of two huge spheres, and their waves. This is supported by the fact that it has become known our mother-galaxy is actually Pringle shaped. Think about it the next time you treat yourself with some seasoned flour.

Knot at a Point

Which smoothly leads us to some new picture of – point-like-stuff. Unsurprisingly, it's all relative. This time it is not relativity of velocity, or gravity, that I still need to elaborate some on. We'll explore **Size-Relativity** in this section, and I don't mean *Length Contraction*. So now that we have some imaginable physicality of what's going on in and within space (the AetherFluid); we can envision some dynamics of what are point-like-particles (the interactions' events). The AetherFluid is the matter that really matters. That is because, all matter is made of it, including that which quickly disappears unless entangled (the computed-particles from subatomic collisions in the LHC data). The DSE shows that. It might even be said, that in the only agreed upon, by-all-physicists, 'mysterious' version of the DSE (Which-Way), the following is the only real mysterious fact. A measurement near the slits makes the wavy nature of particles disappear, when these become (merely) point-like-stuff that is logged as (merely) data. Waves of stuff, from small massless yet energetic photons to some much more massive structured-atoms, on the lowest level of matter, are – Potentials. Basic interactions, the most fundamental of events found in nature, these are bringing-about them – Actuals. Data gets entangled, and entanglements bring-about some additional Actuals, that themselves, get entangled, and so-on-and-on. It's like writing a book, whether it's a novel or nonfiction-popular-pseudo-science. Every additional word (or fact) written, forces some boundaries, and even meaning(s), over the rest of them. Even on sentences that were previously written. Nature is the same. Data and information flows back-and-forth, just like entanglements. This means that existence can be (somehow) seen as (analogous to) some book being written in a single stroke, and that every word being written is in-correlation and in-sync with all that came before it, and all that shall follow. Reality's duality in not-only of waves & points, but also of time. On the one hand, there are events of interactions; on the other, it's all – an event of being.

The image I wish to convey into your fine imagination, is of a – **Knot at a Point**. I can't say that I, with my own set of (physical) eyes, have seen such a knot; but, that is the only image logically-arising in my mind when I melt all that was surveyed here, so far. Once more we are at this point where I put a 'disclaimer' on the next couple of paragraphs and thoughts, informing I'm about to take a risk, and may sound like an imbecile, again. But who cares what others think? Social Relativity has never been on my list of cared-for things. With our leading scientists, physicists, and 'thinkers', unable to agree on reality's fundamentals, a door is open for curious people to offer yet another alternative. Above all, one of the best nature's philosophers, David Joseph Bohm, did just that when he tried to present an alternative to the blundering paradigm of 'WF-Collapse', with a view of *Waves Piloting Particles*. Yet Bohr, a Founder, was able to get a hold on physicists' minds, persuade them there is and there can only really be (pre-measurement) Information on Potentials, that (due to nature's laziness) becomes Data of Actuals only when observed-at, interacted-with. This also holds when waves collide, which is the-same physical event.

So, here goes. Physically speaking – **Space Is All There Is**. The able-to-condense AetherFluid, I mean. The geeks can call it if they wish – *Hardware*. And ask Shlomo Barak, he will gladly explain. According to his vast research and little-nobody-me, everything, all matter, arises from space-deformations; that, themselves result from colliding-waves of the AetherFluid, the 'cosmic-ocean', itself. These waves, just a very-clear derivative of a spherical cosmos, are what enables the creation of those point-like-particles, and planets or stars. Even the (previously thought of as flat) galaxies, and maybe our-entire-cosmos as well, are results of two other (one-level-up) colliding spheres. That's – Size Relativity. A bees' hive seems round from the outside, but when probed inside it is constructed of straight (and flat) hexagonal lines. Nonetheless, when understanding the process via which these hexagons come-to-be, one realizes they are spheres being squashed-around till flat-lines and point-likes are becoming visible (to our human scale of seeing). While thinking in the manner described here, and when adding to-the-mix the fact that nothing in nature stands-still, an image of **Spiraling-Knots** surfaces. Well, at-least for me.

Waves of all sizes and energies are common in the AetherFluid, and obvious. What is less of an obviosity are results of works like of Shlomo, or my own psychedelic-visions and mundane-thoughts. Matter 'pops-out' of space itself, if **Streams in the AetherFluid** are colliding in certain locations, condensing (enough) the AetherFluid by spiraling and swirling it, at-a-point. Then, shapes, and shapes' shapes, are formed. The main reason for the stable (on-average) formation of anything, is the consequence of entanglements and time's creation. More on that in the chapter after the next. When the AetherFluid becomes (relatively) condensed around a point, for example, and photons of light travel through that AetherFluid's condensation, the proper conditions for the rise of matter are formed, thus matter come-to-be. We are starting to see how light & space & time, are quite united.

In 2019's end, a couple of months before humanity has entered a new era of (new) ignorance and disbelief in data and/or analytics, an interesting paper was published with the appealing name – *"Decay of a Quantum Knot"*. Interesting, right? But how do they define a Quantum Knot? (It is) "a quantum object that acts like a traveling wave that keeps rolling forward at a constant speed without losing its shape." Sounds just like a point-like-particle, just multi-dimensional. And if I'm asked (which I'm still not), I would say that the many dimensions of a knot might-be, are, I believe, just the best explanation for the reality of Hidden Variables (HV). A model, a man-made-model, either of math or of decision-trees or neural-networks, any 'Model of Everything', must be subject-to HV. They teach that in the first lesson of Econometrics – only God knows (or is) the full-model.

Chapter 7

Non-Matters

I plea. Do your research. Into yourself. Please.

Evidently, I did all I could to avoid dealing with the matter of (real) non-matters. Till now, where it is time for us to dig there some. We have wrestled with the idea of the stuff that we can see or hear or touch, that way or the other. It is clearly an event, an action of an interaction, resulting in a 'knotted-dance' of spheres, on all scales. But, behind every action there's some energy pushing stuff around. And let's not forget of the 'Dark' one, as-close-as-possible a concept is similar to that of – God(s). "But come-on," you have had enough, "what is Energy?" The term is derived from the Greek word – Energeia (which means an operation or activity). No surprise there, right? In its linguistic-sense, energy is the enabler of activities; and, we know that activities' events are all that matter is, or is doing. Any physics textbook will tell you that energy is the capacity to do work; then, it will usually go-on to explain that work is the action of moving something against a force. And a force? A force is what acts on bodies. The last couple sentences, yet-again, prove the stigmas I hold of physicists, being horrible with words and-all; and, as told, sentences will suddenly feel quite circular once we'll get to the subject of – Energy. We simply have no clear or agreed-upon idea here, as well. Yes, some facts-of-life we accept at face-value.

In the following sub-chapter, I will go-on a venture. You will see. But first, let's very quickly see what Leibniz had to say of energy, and he said-much, so forgive my quickness with it. He had developed a language for everything, since those were the days of The Founders, which makes it extremely hard to follow his line of thought and ideas. But something most-relevant is worth mentioning here, a vision of his, there is just-no-reason for me not-to-use. A force, according to this finest of philosophers of nature, is a principle of activity and change. And, a cause of change must be there, which leads to the following crucially-important notion in Leibniz' theory of forces. There are Derivative Forces (DFs), that are directly responsible for some of the most basic physical properties of bodies, of movements mostly; but, still, DFs are derivatives, not originals. Thus, the basic physical properties of matter depend on the Primitive Forces (PFs) of substance. He suggested a distinction between a 'vis-viva' ("living force") and a 'vis-mortua' ("dead force"), that later became known as *Kinetic Energy* (1829) and *Potential Energy* (1853), respectively, which I'll elaborate-on here below. As always, when things become interesting, somehow, don't know how-come (statistically it is impossible); but, again, we, I, need to mention Einstein.

Einstein's Energy

Energy, like Consciousness, is not a directly observable quantity. Still, many crown them both as the most-fundamental of reality's fundamentals. Very-high-level speaking, there are uniquely two core-manifestations of energy – Potential Energy and Kinetic Energy. Potential energy is stored (or of-pressure) energy; while kinetic energy is that of a motion. Therefore, the full equation for energy's 'quantity', is: $E^2 = (MC^2)^2 + (PC)^2$. This accounts for all manifestations of energy, and not-only for 'mass-at-rest', with a zero kinetic-energy (where P=0). Stored energy (MC^2) can also be seen as one-of-two types. But I am telling you upfront, there's only a single kind of energy, on two different scales of manifestation. It is either the energy that's naturally stored in matter, like the one that was unleashed on Hiroshima and Nagasaki; or, energy stored from an external source, like a mechanical clock where the energy is kept in its winded-up mainspring. Still, one thing is completely certain and simply just-must-be. For physicists to have models of physics – energy can neither be created nor destroyed, but only changed from form to form. This is (dually) known as – *Conservation of Energy*, or – *The First Law of Thermodynamics*. Heat, for example, is the energy that's transferred between objects at different temperatures, always flowing from high to low. In your freezer, heat is being 'sucked-out' of things, and it's not the cold that penetrates the foods and beverages you preserve for later consumption. But, that is entirely not what I wished to discuss here. It's not the venture I've promised. Not one bit.

Unsurprisingly, it was Einstein, with his Special Relativity (SR), who reshaped our understanding of energy. At-the-least, he had made energy a-bit more comprehensible; and directly, mathematically, relating to mass via the speed of light. Certainly enough, he made it even-more circularly-reasoned. This elusive concept, just like consciousness for those who have been researching the Vedas, has been buffaloing the scientists of the West for millennia. And, just as in the case of consciousness, any sane-human won't be able to deny the existence of energy, but, will find it hard to directly enlighten. That's the reason why it always seems as-if some other phenomena are required for the task of describing the physical presence and manifestation of energy. Again, very much similar to consciousness, which shall be explored in the next section. "But where is this risky venture you have promised us, Boaz?", some readers grow impatient. So, here goes.

The case of $E=MC^2$ (*Law of Inertia of Energy*), and its vague relation with the death and destruction caused by the *Fat Man* and *Little Boy* (the atomic-bombs that ended WWII and killed or severely-harmed more than a quarter-of-a-million Japanese), must be deeply delved-into here. There is so-much to learn from it, on physics and math and science altogether. Albert Einstein's 'proof' that the $M=E/C^2$ equation (as was originally written in 1905) is 'real', as he did for molecules and photons, was the assumption he had initially made to (circularly) devise it. It became clear to me as a common-Israeli-sunny-day, when I carefully and inquisitively read – "Einstein's Theory of Relativity" by Max Born (1962). Surely, I went back to Einstein's *Annus Mirabilis* ("miracle year" from Latin) papers on energy & mass as well, but since in them the math and reasoning is evidently circular, I could only assume that later works had fixed it. They didn't. In Max's textbook, he states that $M=E/C^2$ is confirming (that) – "This is the amount of inertial mass (M) that must be ascribed to the energy (E) in order that the principle of mechanics, which states that no changes of position can occur without the action of external forces, remains valid." Told you so, circular-reasoning. Albert was logically, within a 'thought-experiment' (called by the Germans – "Gedankenexperiment"), able to connect energy with mass via the speed of light (C). That is his contribution to this intangible issue, which concurrently he was able to sync with time. But, SR is more-of-a theory, and less-of-a model. I know this is no-less-than heresy; but, heresy is chutzpah's cuisine, and chutzpah is needed for science.

The bottom-line of Albert's theory of SR, of mass & energy & light's (math) relation, has become commonly and popularly considered as 'proven' by the *Nuclear Fission*[1] that's behind the atomic-bombs. But, that is simply not true, AKA, it is (more than 60 years of) 'fake-news'. The story of the nuclear fission's discovery (really) began with the detection of the

---

1 The subdivision of a nucleus into two fragments of roughly equal mass, accompanied by the release of energy.

Neutron – one-of-the-two subatomic-particles of atoms' nuclei. Not to be confused with the elementary-particle, the electron's cousin – the Neutrino. In 1932, in England (of course), Sir James Chadwick found the neutron. Shortly thereafter, Enrico Fermi and his Italian associates undertook an extensive investigation into nuclear reactions. Such events are produced by the bombardment of various elements with this uncharged particle. The fact that the neutron is uncharged, not a minus (-) like the Electron nor a positive (+) as is the *Proton*, is what allows it to penetrate the 'hazy-borders' of atoms' clouds-of-electrons. Then, the neutron collides with, and splits, the nucleus of an atom. This split of a nucleus frees more neutrons, that hit other near-by nuclei, all leading to a – *Nuclear Chain Reaction*.

That's the real science of atomic-bombs' nuclear-fission. The only relation it then-had and still-has with SR, with the $E=MC^2$ model, is the following fact. A nucleus of an atom, constructed with protons and neutrons and fenced by electrons, when ruptured by a penetrating neutron, releases the potential (or primitive) energy that binds it together. It will simply be differently-manifested in other forces' forms. *The Strong Force* is the gluing energy, and its quantum is – the Gluon. Considering that we deal with the smartest people of the planet, when it comes to words they just lack any imagination.

To sum-up, there is a gigantic difference between a theory that states: "For atoms' nuclei, for all matter, to be held-together, a-lot of energy is needed"; and, a (math) model that explicitly 'swears': "$E=MC^2$". Einstein had deduced energy in the same fashion that consciousness was, by bright minds who saw reality in its most subtle form. Me, I think it is much more substantiated that consciousness exists, than energy. As human-beings, it is naturally intuitive, the experience of ourselves, in the same manner that inertia is for physicists, only knowing what movement is. But just before we advance to illuminate consciousness, ours and everything's, I wish-not to leave you hanging regarding the '$E=MC^2$ conspiracy'.

It's just inconceivable that the most famous (and the sexiest) equation ever, preached by one of the finest-minds humanity has ever known, is unreal. Firstly, we know that its metaphysics is as real as it gets. Matter is glued by much more energy (or force) than one would guess. Lots of action, and interactions, occur on the sub-atomic level, and much energy is required to 'regulate' it all. How much exactly? Many observations are needed, as you're well aware of. Which leads me directly to the second point I wanted to convey. Once we have realized, and hopefully agreed, that nuclear fission is not (at-all) a direct observational-proof of $E=MC^2$, I challenge you to find its real proof-with-measurements. If you accept this mission, I promise you some interesting insights. Not only that you'll find Einstein's video declaring it was in 1932; it was also in 2016 and recently in 2021, all for the 'first time'; and, you'll also find it was proven after 'heavy', 'heroic', computation. Indeed, just like economists, and more-so statisticians – give us data (and money) and we'll compute any proof your agenda requires. Have no doubt, that is how the Higgs was 'found'; and, that's the best proof that particle physics, that any particle – is ('just') data.

Good Ol' Consciousness

Finally, I sense we are getting somewhere. Hope you feel it as well. In this section, after we've touched on energy, as it is manifested in all-sort-of forces, let's go back to the basics and deal with the-real-deal. Truly, the only fact that we, humans, can be certain of. What else exists, what else we're really aware-of, if not awareness itself? Nothing else we see, besides ourselves seeing. "There is no spoon," you probably heard. Do you think we see photons? Do we see light? There is a physics' paradox hiding there, since seeing is – the processing of photons themselves. Photons deliver data of something else, an interaction, as we have learned; so, the answer to the above question is: No. That's why Serge Haroche received the 2012 Nobel. He devised a device allowing the (indirect) seeing of photons. It is achieved by *Controlling Photons in a Box*, and it's as cute as it sounds. In order to analyze photons, indirectly, and not the data they carry, we examine a box where they're trapped.

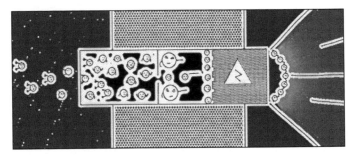

But, back to the non-matter that really-matters. I firmly believe that we've played in the physical matter's mud enough, haven't we? So, please, allow me to have some fun; many thanks in advance. If you are still here – you will probably enjoy what comes-next. Consciousness was where my journey into physics initiated. In hindsight, I thought some-more physical evidence was needed for the philosophy and metaphysics I have (felt) I mastered. But, again, nothing is more physically-evident than our consciousness itself; because, and we must agree here, all sensations are produced within it, even 'made of it'. This rationality can be labeled as – (New Age's) **Qualia**. Chopra and Kafatos are central in spreading this idea. I think it's great, but it does not speak my language. It is missing a model. And me? I am a dumb-motherfucker just like the rest of them – I need a model.

There are (at-least) 16 'academically accepted' models of consciousness. Only lately it was attempted to classify them all, in the journal – "Neuroscience of Consciousness' Special Issue: Consciousness Science and Its Theories" (2021). Exciting times for reality, and maybe for truth, too. Now, a confession. The main reason I slammed through these Dark-Variables and Spiraling-Knots, and swam in AetherFluid of Potentials' Information that collide into

Actuals' Data, was that we'll have some common language to later delve and explore Consciousness. Speaking in these 16 models' language, and those surveying them, would cause us both pain. For example, they compare models by 3 dimensions of explanation: Mode, Mechanisms, and Target. I bet it does not entice you, in-case you are not a professional 'mind-researcher'. So, let's do it my way, with the language we have together established. Let's speak of consciousness in the language of physics. There's a good reason to do so. We live in a physical world, and know nothing of any other.

I should start with a question. Who, or which, is more conscious – you or our Sun? I'm sure you now in-response ask: "But how do you define and quantify, consciousness?" A definition was provided in the Definitions' sub-chapter of the Intro. Let's start with the traditional concept of consciousness, as it was philosophically derived through works that are based on the collections of scriptures and lectures, known as – the Vedas. This collection represents the most exhaustive research into this (even more-so than energy's) elusiveness. Let's examine its main philosophical and logical maneuvers. For simplicity, and since it was my own doorway into Vedanta, I will explore it using Advaita's logics. An Advaitist-Vedantist would openly declare that there's only an ocean, and its (colliding into point-likes) waves are ephemeral, constantly changing, certainly not real. The only reality is the ocean, from which the various waves emanate. As a physicist, such a thinker would clearly follow the works of researchers like Shlomo Barak, and shall accept my complementary metaphysics. However, this is only its stance on matter. Jñāna Yoga, that I recommend reading every word-of-it (it's all available online), is the finest metaphysical summation of Advaita-Vedanta. It was perfectly preached by Vivekananda till his death (1902), at the age of 39, in all his attempts to reshape Advaita as a universal religion. It is scientific in nature, and speaking in a language that even contemporary physicists should be able to digest. Unquestionably, it should not deter them as the word "religion" does.

"Dvaita" means duality, and "Advaita" means non-duality. In simple terms, Advaita means an absemse of duality between subject and object. At its basis, it's entirely in-line with Bohr's (or Bohr is with its) notion that – an Observed requires an Observer. For data, an accepted fact of an event (of an interaction), to 'pop' from an information's hazy-wave, an attention of some recipient(s) is-a-must. Yes, physics-wise, Bohr and his disciples were as Vedantists as it gets. But, the extra-step towards the illumination of consciousness was never taken by them. I guess it lacked the math needed, or, their attention was simply pointed elsewhere. But, lucky-us, there have been and are Vedantist walking among-us. The cornerstone ploy we offer, rolls as follows. Any observed, to-be, requires an act of observation; and, an effect requires a cause. These two parts of a sentence are consistent.

Now, and please pay attention here, thinking (that is the processing of data, as we've seen), our human-thinking (yet not-only), is also subject-to this reality of observed and observing, just like anything that exists. Witnessing (or observing) ourselves thinking, or **Thinking of Thinking**, is the best physics description one can provide for Consciousness.

It is the "Essence of Vedanta" (a very nice book by Brian Hodgkinson from 2006), and I think its logic is sound. There is no recursion ad-infinitum of (or for) thinking. Only just a single-step to a higher-order of thinking is possible. There is no, and none-can rationalized, witnessing of the thinking of thinking. None-can observe their thinking's thinking. Only 'one-level' higher of this thinking's recursion is possible. This reasoning, of our True-Self, seen as a 'quantized case of an Ultimate Reality' (Brahman, as it is traditionally called), if combined with the attitude of ephemeral-waves-and-points towards matter, easily leads to *Religious Fundamentalism*. Neo-Vedanta was supposed to remedy this extremism by offering a finer view of this nihilistic philosophy, which openly denounces matters of the 'living-world' as merely an illusion (called – **Māyā**). Yes, a common Vedantist holds just the opposite world-view than that of an average Baby-Boomer. We put the weight of existence on the shoulders of philosophy-of-consciousness, rather than on those of physical-empiricism.

Post-Neo-Vedanta

Which leads us to the latest, most modern, evolution of this ancient philosophy, devised solely to speak both to empiricists and meta-physicists. Denunciation of physical-reality, as Advaita concludes-in, as it has influenced me, simply won't allow Vedanta a place in the hearts of Western-humans. "**The Complete Works of Swami Vivekananda**" (found online), are the summation of Vivekananda's effort to deliver a universal, socially-moral, religion. One that relies-on and completely accepts the unity that's at the root of Advaita-Vedanta. Much later it had received the label – Neo-Vedanta. As previously declared, I am an Advaitist turned

Neo-Vedantist, and here you should be asking: "What is it that separates the Neo from the (original) Advaita?" Between these two there's a complete agreement on all-accounts, but one. Quickly I'll put a fat-finger on their intersection. The foundation of agreement these philosophies share, is of consciousness' fundamentality and primality. They both agree that consciousness is existence's 'energy', and cause, and was as-such prior to the Big-Bang-Inflation, and shall remain-so even when we're gone. It's always the same, an unchanged (ultimate) awareness, and/or, observer of existence.

Now, Advaita vs Neo must be clarified; primarily since I intend to further evolve the latter into an even-more-solid version of Vedanta, one that embodies the facts of physics. In both philosophies, Consciousness is synonymous with 'God', and, as a succession, we are all just a part (**Ātman**) of an ultimate reality. Ātman is the quantum of Brahman, they both agree; but, the clear-disagreement lies in their attitude towards physical matters and other mundane structures, i.e., towards – Māyā. While Advaita sees all shapes and forms as illusionary, as stuff that can only drag one away from reality and truth, the Neo views all physicality as God who manifested into space-time, and, cause and effect. Unlike the Advaita, which accepts no-form(s) of God(s) besides consciousness, eternally existing and with no relation to physical substances and/or the laws of physics, the Neo-Vedantist sees God in everything, which allows one the finding of **Interest** in (physical) living. The devoted Advaitist finds himself in a state of a **Refusal-To-Do**, while Neo-Vedantists are in constant search of what it is that is interesting to them; what it is that their Self, or Soul, enjoys-the-most

(**Ānanda**). To summarize, Neo-Vedanta rationalizes the 'magic' in life, by viewing it as a manifestation of God, and not just a neglectable damn-well-lie. Yet, it still gives consciousness the attribute of being eternal, infinitum, which I think is wrong.

Within Vedanta's next evolution, let us call it ('Quantum-', or, 'Aether-Vedanta', or, maybe, 'Dark-Vedanta', or, most-simply) – **VeData**[1]; Consciousness is not, only, merely, of this physical world; it directly interrelates with (processes of) the processing of – Data. But, swiftly let us take a step back. An insight that I must convey here, is that humans are religious-creatures. "What does it mean?", some non-religious-readers now think; and, "How do you define – religion?", some also might add. Religion is (in complete contrast with – 'Established Religion') defined as the Filter, or Grid, even Model, via-which data is interpreted. Humans, we, have evolved to observe and absorb data through contexts. Using our language – with entanglements. This is the most scientific, dry, and objective definition of – religion. When it is viewed as such, one cannot escape the conclusion that all humans, unlike dogs, are religious. We simply can't do without models. Once I asked a girl for her religion. She's a

---

1 There is a firm called VeData, so, I ask them up-front to pay me for marketing, or, at-least not sue me for rights.

Brit, and as secular as Brits come. After I gave her the above definition, and some-time to think-some, she came back and declared: "My religion is my Mother." Right there-and-then I knew she understood me, and even learned something new of herself. Have no doubt, if you are a human – you are (this kind of) religious.

The additional truth, the real-evolution, I offer to the lively philosophies of the Vedas, is a strictly-physical perception of consciousness. Neo-Vedanta separated from Advaita by removing from the AetherFluid and all matters the label of – illusion. Yet, the god-like traits it attributed consciousness remained as traditionally assumed. Consciousness is still regarded as a representation of something that was before, and that after will be, with no beginning nor end. Infinity. We're completing a full circle now, going back to page-1. In our language, Neo and Advaita-Vedanta, claim that consciousness is something that was-as-is also in the days of **0**; when days-were-not, when simply nothing-was. This is a grave mistake, one it is time to correct. Naming the time took for the Big-Bang-Inflation to 'build' the universe; is as-ridiculous-as attributing some phenomenon of this world to any that was also in a time with simply nothing at all. Speaking of such inconceivableness, is even more misleading than the stories of dancing blackholes in far-far-away galaxies.

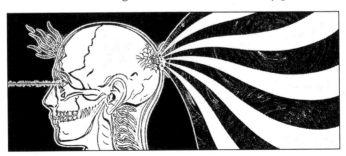

The condensed metaphysics below may remove the 'magic' from consciousness, and from psilocybin (magic) mushrooms as well. All such and relating phenomena will 'simply' be modeled, very-macro-level-speaking, as follows. An act of observing, attention giving, leads to the creation-by-interaction, and absorbent of, data; then, the processing of data, essentially the witnessing-of, gives rise to entanglements, manifested as memories (of the past) and (of the future) predictions; these themselves, synchronously and immediately, are followed-by an update of knowledge, of core-models that are constantly producing predictions, and updated. This entire obliquity can be called, or even, is – consciousness. That is the reason why consciousness is of-this-realm, only, and not magical nor infinite. There's zero-logics in attributing it with a beyond-this-cosmos existence. The principal motivation to accept this argument should be the following understanding. Our entire existence of relatives (either of matter or abstracts), is in-itself relative to which that was in 'pre-existence', during – **Relative Nothingness**. But, this relativity is not an **Absolute** (Nothingness); and, as it is misleading to speak of the time took for Inflation, it is equally wrong to speak of a prior-consciousness. No, it was purely some-kind-of – **An Elseness**.

CHAPTER 8

Levels of Dark Variables

Because there are. Levels to this. Baby.

The End is near. No, I'm not getting dramatic; and no, it has nothing to do with COVID. We are at the 9th chapter out of 10 (don't you forget it started at 0, as everything does), and the last chapter of exploring consciousness was awfully unsatisfying (yes, I know). But, if you are reading these very-words, it can only mean you trust me by now, and know there is zero-probability (it is not in the information of this book's WF) that I'll leave such a subject unconcluded. Even my only novel, which was left unconcluded, was essentially much decisive. So, with this promise, here's what's to follow in the final two chapters. In the coming sections I'll deepen into and summarize the two subjects that were at the focus of this work. First will be the AetherFluid. We'll see how its density can be (also) seen as onions' scale-leaves, and to where such a view of the cosmos leads. This will directly, and finally, bring us to the subject of Time, which long-ago you were assured an elaboration on. Once I'll be done with the presumptuous act of defying it, I'll continue my brazenness into the second main subject that was covered; which'll again be clarified by making an analogy between our own Levels of Consciousness and 'The One Eye of The God(s)'[1].

---

1   "Eye of the Gods" (2004) is also a Nigerian movie about a princess who dislikes her (picked-by-an-oracle) husband.

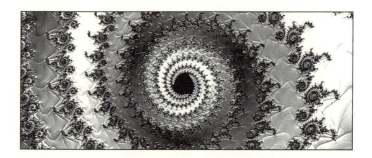

But quickly now, just a couple of sentences about the knots and flatness and points that were previously disjointedly mentioned. It's also a recap prior to the rest of this chapter. *Fractals* are found all over nature. In his (2013) *TEDx* talk about fractals, Michael Frame said: "A fractal description of an object is a story about how it grows. Fractals remind us that science has a narrative component, that we too often ignore... (and that) stories are important." This assertion references Benoit Mandelbrot, a Jewish-Lithuanian-American who died in 2010 and was the first to coin the term – "Fractal". Quite interestingly, the 'Father of Fractals', as he's universally known, considered himself a storyteller, above all. Personally, I cannot agree more. Told you already, math is not the language of existence. Language – is the only language that is capable to encompass the most aspects of reality. So now, in plain English, I'll present a macro-to-micro (fractal-like) image of the cosmos. Higher-levels of knots' knots, or, a universe of **Finitely Many Touching Circles**, seems as a fitting model, or a story, for our physical existence. The very evident and much needed sphericity, allowing micro-point-like-particles and seemingly-flat-macro-structures, can easily be imagined stemming from such a universe of spheres. Thus, we are led to picture the world as an indivisible, though flexible, forever changing and becoming, single entity.

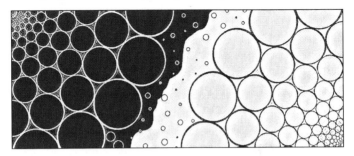

An Onion Universe

Quite surprising, in the book "The New Cosmic Onion: Quarks and the Nature of the Universe" by Frank Close (2007), there are no onions nor physics' analogies with onions. 'Nonions' (sorry, had to write that). It is a great book on QM and physics, but, stupid-me, I was expecting some Theory of Everything (ToE) with onion-images. The reason I didn't, I believe, is rooted in the fact that space-matter-density was just a step-too-much for our past-physicists. But not for me, and not for several established (real) physicists. So, expect some onions here. In 1924, Edwin Hubble showed that many-many galaxies exist. This discovery indicated the universe is greater than was imagined. It was followed by another astonishing discovery (1929), that these galaxies are all rushing-away from one another. And in 1998, struck the latest astonishment, of the changing speed of that rushing-away, and Dark Energy (DE) was declared – 'God'. In this section I wish to explore yet another possible effect of DE on the AetherFluid (AF). If we accept the possibility that the AF can vary in density, via streams and waves that build into

AF's knots of knots, it may solve some issues in cosmology, and will probably make no-need for any Dark Matter (DM).

So far, I did not say it explicitly, but I was hinting at this on more than one occasion. DE should not be credited with the expansion of the universe, 'only'. I see no reason why not recognize DE for everything that exists. If space is all there is, and all matter, from 'big' to 'small', all is merely (entangled) deformations of the AF; then, why just the expansion of the AF should be the result of some mysterious DE dynamics? Why not all that arises from it, in events of colliding waves, is caused by DE? Already DE is said to be the force behind (roughly) 68% of existence. In my metaphysics I only assign the 27% of DM to the AF as-well. By doing so, I unify the two 'darks' of cosmology – into one AF's dynamics. When we've explored the concept of energy, and learned it's a placeholder for describing some movement, some inertia, and combined it with advanced ideas of spacetime matter; it seemed as-if everything can be explained with the AF and its waves (as Barak so-did).

Still, that was only to explain elementary particles, as Shlomo's GDM focused on; and their structures of entanglements, as I have illustrated in an attempt to explain the higher structures of actuals (of – data). But, there is another mean to (ab)use these two-in-one assumptions, that – everything is AF & DE. See, if we tolerate that the AF deforms into localized knots-like interactions, that are AF's colliding waves, we should also allow it to condense, and create some **Cosmic Paths**. Previously, these were mentioned here as the Geodesics, paths of similar 'gravity' that photons surf. But, there's more to it. I think there is another derivative to this density, to AF's condensability, if we imagine it can establish some quanta of cosmic-onion's scale-leaves. This will mean that not only 'Cosmic Paths' are formed by the AF's condensation; also, **Cosmic Lumps** are formed, which can kill the concept of DM. Now, let's examine this idea; and right after, see how it clarifies – Time.

The cosmos, the universe, is an AF's ocean. Particles are springing from it, as swirling knots of waves and streams, and get-structured due to entanglements. But there are levels to these knots within knots, baby. It's a fact that galaxies behave as if they are swirling in a unitary manner; and, when observed-at from very far, may look just like some atoms. Remember, atoms are entangled, in-sync, hazes of particles. Do you think we have ever seen an atom? The very short answer is: No. We've seen a point, maybe, or inferred the existence of something by (ab)using data from indirect interactions, and then computing (within models' frameworks) how things might look-like from up-close. We are surely not seeing the quantum and the cosmos as we see things within our onion's scale-leaf. Which leads to where I was obliquely roaming. It was the earlier model of the atom, the (1913) *Rutherford–Bohr Solar System Atom* (which was deemed wrong), that I think exhibits an accurate 'metaphysicality'. These two were, surely, some of the-best-of-scientists ever.

The AF is condensing. On the lowest of (all) levels, where there's (relatively) a-lot of energy, as in the nuclei of some heavy atoms, point-like-quanta are being formed; but, on the (relatively) higher levels, a-kind-of scale-like-leaves of similar density (not of similar 'gravitational force'), come to be. These cosmic scale-like-leaves can be analogized to the clouds of electrons within atoms, if we realize the analogy can be labeled as merely one of – **Borders**. Let's now quickly story-tell the tale of our solar system, that is said to be bound (or held) by gravity. Bottom-line, we have no idea what gravity is, since physicists only know of quanta; and, we have-not yet found any such for the 'force of gravity', and will most-probably never find one (nor for space itself, as some of them search for). We live in a bordered universe; made of bordered galaxies; themselves formed by bordered solar-systems; themselves are structures of entangled matters, surfing in bordered-paths; that, as we've quite deeply explored, fundamentally are made of knotted-like-events; also bonded-together by a *Strong Force*, and, obviously, found only within bordered-locations.

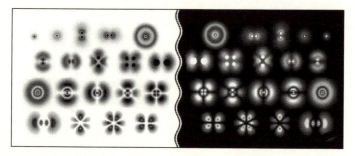

This very-short-story I just told (must admit – never liked nor wrote short-stories), is how the (fractal) tale of our physicality is grasped in my mind. It also explains where

Why Science and Psychedelics Go Hand-In-Hand | 161

and why this sight of matter is essentially the opposite of that of everyone-else, more-or-less. The significant majority of the best physicists (not-including Shlomo Barak), believe that matter is distinguishable and completely separate from space. That is one-hell-of-a space, if its 'gravitational' (nobody knows what's gravity) waves can squeeze the Earth, without being some integral part of it. Furthermore, this is also one-hell-of-an Earth, said to be curving and dragging the AF around it – largely accepted as the cause of gravity. (Did I tell you already that nobody knows what is gravity?) The Earth does-not drag space, nor it is squeezed by its surrounding-and-only-touching AF. Our amazing yet hurting planet is at the heart of one-hell-of-a 'Cosmic Tub-Ring'¹, made of, or by, an AF's condensation of knotted streams and waves. It is just the same energy ('dark' no-more), the one said to be holding those atoms as-one; that holds solar-systems, and galaxies and their clusters, all-together. What's most special, is how higher levels of AF's condensation give rise to both point-particles, and create cosmic-scale-leaves for planets to surf. It is all just – Time.

---

1   I don't mean the physics-inspired, quite-psychedelic, Chicago band. Their "The Great Filter" (2007) is amazing.

The Metric of Time

It's time for – Time. Here, the density of the AF shall be knotted with it; or, even more precisely, we'll see again how time is matter, and vice-versa. And, since Mass is Energy's manifestation, a relation between energy and time will be formed as a byproduct. Some of it was previously glossed-over, when Feynman Diagrams for particles' interaction, or the point-like-derivative, were cited. We saw how a metaphysics (and physics) in the likes of a Quantum Handshake, where the idea of time as principally a Least Principle Action, and vice-versa, is reasonable. We know that matter is just energy condensing the AF itself, becoming denser into entangled particles. "And Photons," I hope some ask, "the data's bearers of those interactions' events, how do these 'knots-tellers' fit in your story?"

Well, Photons are a huge part of it. You see, as the fastest particles-out-there (maybe Neutrinos stand some chance), and the only thing-out-there that is massless; the role of photons (data) in the process of the (relativistic) creation of time, and time's metrification, is pivotal. Barak provided some math for such an idea, without shining a spotlight on it. Just by reading the $E=MC^2$, mass and energy sameness 'proof', mainly in its original form, it is kind-of straightforward that the speed of light, squared (C^2), is the regulator between the two. C, which is almost 300 million meters per second, is the – Ruler of Time (with the dual-meaning here). Just a brief two side-notes regarding – C. First, the velocity above is only for photons in a 'vacuum' (which we should know there's no such thing); and secondly, the speed of light for energy is what an interest-rate and a price-index are for money. It, they, make it quantitatively 'real'. Lastly, just by the way, the speed of light was never directly measured, only its on-average velocity was truly observed. Yes, yes, some more on-averages, and more Dark Variables. Surprised? Well, it's time to grow-up. Now, let's see how those bearers of data and the density of the AF create the metric of - real-time.

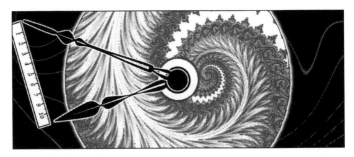

The following maneuver is not so different than Einstein's endeavor with GR, only the story here is a-bit different. And there's no-math, just words. Physics still did not explain nor agreed-upon what time is. Subsequently, even the finest physicists get blundered with this concept. The inspiring Carlo Rovelli, for example, a great storyteller (which I think makes him an even better scientist), wrote this in his "The Order of Time" (2017): "The Earth is a large mass and (it) slows down time in its vicinity. It does so more in the plains and less in the mountains, because the plains are closer to it." First, a clarification. It is well observed that a clock that is either on-the-move, or located closer to a higher energy's condensation, will literally tick less than a clock-at-rest in deep-space. Secondly, the above sentence by Carlo really bothered me, trying to make sense of it while acknowledging that there's politics to it. If someone like him feels-free to publicly state such a statement, it can only mean this is the now's paradigm. No other explanation.

Do you visualize the insinuation of Rovelli's sentence? He and his mates think that somewhere at the 'very-center' of our planet, not a-bit 'right' nor little 'left', there's a point (yes, another point) where time flows the slowest. But they rationalize it the wrong way around! Using the language that established here, at that point-location the AF is densest, because it got most-condensed by the forces of knotted streams and waves, themselves a manifestation of the AF's (core) energy. This might sound quite New-Ageish, but it is not. Yes, it's a metaphysics; but, most things are before they are realized. I'm not saying this is true nor that it's real. Remember Bohm, who's making a grand comeback. Time, on the other hand, is not even metaphysics. At its current status it's an illusion, a (real) Māyā. This is not an understood-idea; completely half-baked it is; and quite neglected, I believe.

Building on the vision of an 'Onion Universe', of AF's varying-density, while clinging to (Einstein's logic of) a constant (fastest) speed of (massless) light, a new understanding of time is conceivably imagined. Time, so it can be appeared, is the metric of the Energy that's binding an AF density. The closest term physicists have for it is – "The Strong Force". This we've already fully-knotted with particles' energy, when we saw how all can spring from waves of AF. Then, when this is combined with the realization that on the quantum level there is no space nor time, we were forced to view matter as events of interactions, that are time itself. Conclusively, this entire maneuver establishes some-physical-cycle, of how Energy in Space creates Mass and Time. Now, I'll try and explain, using the above, the two accepted types of Time Dilation, and add another, for – Levels of Entanglement.

Now, please, imagine three Swiss-clocks, the golden-standard of (traditional) time measurement. These three are in-complete-sync, effectively the-same clock times-three. Clock-1 is stationary near you; clock-2 is also as-near, but was placed on a toy-train that moves in circles; and, clock-3 is somewhere on one of the shores of the Dead-Sea, the lowest land on Earth. One thing we know for sure, clock-1 will tick (and 'grow-old') the fastest. The two others will be subject to some level of time-dilation, relative to clock-1, of course. It's safe to assume that clock-3 will be ticking the slowest, because the Strong Force is, well, (probably) the strongest. Why is that so, and how does this (New-Ageish, yet another) 'Theory of Everything' I have just portrayed, explains these observable facts?

Time is truly the most relative aspect of reality. It is the 'last-in-line' to be determined. Just like the **Plug-Number** too-many accountants (ab)use for balancing a patchy quarterly earnings-report. Why? Because it's (relatively) merely-existing. Soon more on that. Now I remind you of Einstein's two-photons, bouncing between two mirrors, 'counting' the 'time' that had passed as the bouncing itself, as the interactions. That was yet another reason why we've concluded that, according to this view of time, time is events of interactions. Me, I don't completely accept the (Pythagorean) bouncing-photons, but understand this is how they think of it. Another thing physicists understand, is the Uncertainty Principle. The Uncertainty Principle (UP) has two manifestations, as most things in physics (always two things existing in-relation to – Nothing). A quantum-event, a least-action, can never be simultaneously determined in both its velocity and location, or, its energy or duration. The best-of-physicists declare (and explain why) there is no time, and know it's a physics' textbooks fact – Time and Energy are entangled more heavily, than Mass is with Inertia.

So, what's needed here and now, is to present a theory, some high-level metaphysics, that can explain the very observable time dilation of velocity and 'gravity', by utilizing the vision of photons in AF's congestions (waves that interact into facts of events – into data). The energy holding matter together (and its higher structures), which Entanglement is a manifestation of, is more powerful than any force that pushes stuff from here to there, and everywhere. Kinetic (for derivate inertia) and Potential (for primal binding) Energies, macro-differentiate them. Therefore, the time dilation due to density of AF is way-more significant, and obligatory, than that of matter's inertia. It's also cool there are two types of Time Dilation, matching these two types of energy's manifestation. It makes sense.

Back to our Swiss-clocks. On any of the shores of the Dead-Sea, the AF is denser than on top of Mount Everest, thus photons move slower between the mirrors, and less time is created. Just like the photons on the train, experiencing less interactions' events. Light is clearly moving slower through water, or glass, thus it is most assuredly moving slower through the AF itself (in the case one accepts the comeback of some Aether). But it is not really moving slower, because slower means there's something called – Time. But there isn't. What there is, is simply a local regulator, the AF's density, making sure that data, photons, flow uniformly. But somewhere else, close to some blackhole, the AF is so dense, that the cosmic onion-scale-leaf is so (relatively) thick, that photons get stuck-in 'for-ages' (yet not necessarily get completely sucked to its very-center). But, for some reason, there seems to be a tradeoff. If you'd move at the speed of light, the strong force, primal energy, will simply leave you and go somewhere else it's needed. You know, just as nature loves to do, being so-lazy and all. The faster something moves, the more inertia it posses, the less Potential Energy it encompasses. I (can only) guess it 'feels' less needed, since (I imagine) the AF, like ocean-water with a whale-shark swimming-in, flows around matter when it moves-within. Einstein was close, yet not accurate, I think. When matter is being pushed around inside the AF, like billiard balls on a table, the inertia and the interaction with the AF aids with the task of its condensation. This also self-explains the known phenomenon of Length Contraction, shape-change, which is in-some-agreement with the above story.

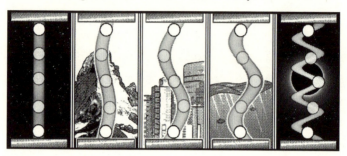

I wish to quickly conclude this section on time; of photons bouncing in trains, and in space and on mountains, and in the ocean or near blackholes; with yet another metaphor. This is – of falling dominoes. Just like most humans, each sees only itself, and just a-slight to the right and a-bit to the left. In the 'Domino Effect' of matter, each brick sees only its neighboring front and back; thus, it is existing in-between them, only. The denser it gets, the more knots of streams and waves of AF are at some domino's location, the more time it'll relatively take for it to fall. Entanglement (a manifestation of energy and information), I think, I believe, is what keeps the bricks as structured-particles, and even those particles (events of knotted-interactions) themselves; and, on-the-highest-of-levels, this what makes them fall into a bigger-larger coherent shape. This is as fractally as a process may get. On the-lowest-of-levels, not-relatively, thus quite absolutely, things might seem as uncertain, subject to (both higher-and-lower-levels of) Dark Variables; but, on the-highest-of-levels, from an **Ultimate (Self) Observers'** perspective, non-is-uncertain. What's now needed is to find the minimum action in nature, some least physical event, and that most-very-basic 'Quantum Domino' – only it can and shall serve as the real measure of a real 'tick of time'.

Gods' Eye

Here comes yet-another Jew, about to profess you what it is that Gods do. No idea what's the deal with us, Israelites...I don't rule-out the possibility it is of a pathological nature... The utterly disproportional percent of Jewish-men mentioned in these pages, and among Nobel laureates, is almost ridiculous. At-least 20% of Nobel Prizes went to those of (some) Jewish blood. Currently there are a-pinch-more than 15-million of us worldwide (still less than our number prior to The Holocaust[1]). Do the math, we are less than 0.2% of humans. So, in-line with my heritage, and what is expected of me, here goes. I wish-not to skip an opportunity to put-in my 2-cents on such a loaded-subject. In a sense, it's kind-of a-must, quite unavoidable. Since (some) 'Post-Neo-Vedanta' was presented, and a (somewhat) physical-model of Consciousness was provided as well, an account for the

---

1 The state-sponsored murder of six million Jews by the Nazi German regime & Co. (in-case you have forgotten).

idea of God(s) should, must even, also be portrayed here. But, just a brief recap of what I call – VeData.

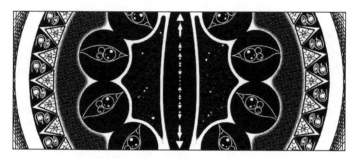

Data is all there is. Do a séance, ask Bohr, we know he will agree. If you'll discuss this with Barak, he'll say that it's all just AF's deformations, caused by AF's waves, which I'm in complete agreement with. But, unlike Shlomo, and unlike too-many physicists, I won't dismiss nor neglect consciousness. This is why, from a mind's perspective, data is all there is. All that our mind does, and it is probably true for all creatures (and organs, and cells, and certainly particles on all levels), is to process data, that way or the other. The mind is not conscious, nor in-it consciousness resides, nor from-it it-arises. The mind is an organ, just like the heart or liver. It's a piece of hardware, only this one is capable of running some very cool *Software*. Every other organ intakes and reacts to data, transferred to it by some medium. Every cell of your body is a mini computer-machine, even though it has no screen. Somehow, explained only by entanglement(s), by some synchronicity, it is able to force all its point-like-particles to manifest the most-amazing-team-work possible. Any coach of a football team prays for such coordination among his (yes, his) players.

Answer me this, or at-least to yourself. Will you be able to view yourself as a cell? It's not like I'm recommending, but having a condensed and thick shrooms & lemon ice-tea, in a clear-skies' night, will make you understand how small we, you, are. If it wasn't clear from the image I've painted, our entire human existence is not even, it's far-less, than a passing of a wave or a bubble of a wave's foam or the dirt on that bubble. It is the reason why time is of no (fundamental) essence. We must be able to comprehend the (relative) inexistence of the human concept of – time. You're not going to be measured by the sum of years you are alive, nor by the number of offspring you leave behind. We are only, solely, being measured by two parameters. Firstly, is by our consciousness' borders, which is our mind's information (or 'WF'). The more we are aware of the possibilities of what can happen – the more real and the closer to reality we are. The second, is by the interactions of data we've produced, and consumed, and to what extent it got entangled. Do you think Albert Einstein is dead? Physically, within the concept of (local) spacetime, in the AetherFluid (that we already know there are entanglements beyond any physical transfer of data via photons), Einstein is more 'alive' than ever. And Adolf Hitler, is the data he had produced and the many ripples he had caused, have they completely calmed?

Time is a quanta's byproduct. And quanta? It's of a relative existence, and vice-versa. When it comes to the-highest-of-levels, the 'Universal WF', we simply have nothing-more to compare-to. Just like the thinking-of-thinking, which is clearly the ceiling of our – Self. Imagine a seed, of your favorite fruit or vegetable, or a flower or a tree. It is information, bordered-possibilities; beyond it being matter, which is data. Can you accept, on the other extreme of existence, that everything is an idea, an image; or, at-most, a dream happening is some higher, the highest, entity; able to process and sync much-more data, all the data that there is? For us, a Quantum Handshake, a 'Least Knot at a Point', is the minimum comprehensible event. For some higher object, interacting with (creating and absorbing) more data; for which we are simply fluctuations and interactions (within borders); we're probably the-minimum-of-events. For the Earth, such a massive mass of entanglements, the 'time' we are on-it is of zero-significance. It got proper proportions. It views us, it only 'cares', if we are – alive or dead. This is what I meant by Size (or Entanglements) Dilation.

If you accept this concept, how do you answer the previous question about the Sun? The answer is obvious, right? The Sun in much, much-more, conscious. What are we, if not a 'WF of genes', evolving and becoming with the processing of data? Most New-Age philosophies assign the human mind creativity in its highest sense. We are told we create our own world. There's much truth there, and I'll touch on it soon, but, I think it's more of a psychological essence than of a physical. Physically speaking, it is clear that nature does not take form, does not 'invest' Potential Energy, if it's not required. This brings me to – **Awareness of Levels of Determining**. If we have realized this reality, that we influence it only by interacting with it, communicating and consuming its data; what makes us humans think there are not levels to it, and higher awareness doesn't lead to a higher influence? It's very hard for me to answer the question – if humanity will find a 'Universal Minimum Grain of Time'. Maybe the photons in a box, or some decay, or some spiral of a quantum-knot, or a handshake – will be such one. I have-no-idea. But, if it will be found, I'm sure it'll be the new 'God' of scientists; and probably, not of-scientists only.

Chapter 9

The End

This is. Probably. Just the beginning.

That's it, the End is here, no longer (only) near. I hope you've enjoyed, got intrigued, and found a couple of subjects you wish to further deepen your understanding in. With all my heart so I hope, since that was the main incentive for this (not-so) humble effort. The physics basically ends here, and the rest-of-it, besides some very-brief summarizations here-and-there, is almost completely of the New-Age's nature. Just a little bit of New-Age never killed nobody. Or did it? Now I'm not sure... But, anyways, it is time to sum-up and make some of what was surveyed and developed here, as advertised, quite practical. But, just a couple of lines beforehand I will go-full-circle back to the 'source', to – Zero. If the physics or the physics of consciousness alone were your points-of-interest, I guess it is time to say goodbye; you're not going to appreciate what's to come now. So – goodbye!

For me, in my thoughts, it is clear that **0** is not a number. It's a digit, indeed, but, more than anything else, zero is a concept, just like – ∞. Infinity, which is accepted as a concept and not a number, is also recognized as not of this physical existence of ours. I think the case-of-the-zero is almost identical. In your bank-account, as with the casual-game's digital-wallet, there's a zero exactly between -99 and 99 ($ or € or £ or Q, or whatever). Sure, an account might be empty, fill-with-void; but, there's still an account at some bank, instead of the real-nothingness that can be imagined as the true-opposite to any physical existence. No-wonder that zero was introduced by the thinkers of India, who (naturally) treated the physical existence as an illusion, and sang against attaching to such a delusion. But, here too, zero has a double-meaning. It indicates an emptiness, and, a non-existence.

This helps me comprehend the fallacy in the misinterpretation of an 'of-this-world' consciousness. Allegedly, it's a manifestation, a remnant, some even claim, of a thing that was 'in-existence' during the 'time' of 'pre-existence', when non-was; manifested now as the Aetherfluid's Energy (and the Information it 'holds') and Mass (that is – Data of Interactions' Events). Speaking, even just the thinking of: "The cause for the universe" (a very hot topic for 'physicelebs' and YouTubers and New-Age writers), of that event that still is, just makes zero-sense. Not logics nor a measurement will ever get one there, to such a conclusion; only some (not-so) sophisticated charlatanism. What we can speak of, if you accept (at least most of) the ideas that were offered here, is what I've presented in the last sub-chapter. Understand this, to (really) Predict the future, not just in-probability, is in-fact to See the full (and accurate) present. Any other 'prediction', outside the ones delivered by an 'All-Seeing-Model', a true Theory of Everything with no Dark Variables; is no-more than some (very) educated-guess, more-no-less. Because existence is an event, singular (like truth); quantized into entangled interactions by those who do-not view it in-complete; an 'Observer' is not just a physical term, it is – Human. As Carlo Rovelli has been trying to explain for 25 years – events are observers, as much as an observer can be.

"Are you sure?", suddenly you are not sure. Wait, what, so-quickly you've forgotten? What are you, physically, fundamentally, really, my dear fellow human? Our body is not at the top of the chart, to say the least. Cockroaches are far superior. Trust me, I broke an ankle a couple of years back, and learned how poorly the human body is designed. After which, by the way, I completely stopped trusting those hospital-doctors, discretely, and began averaging the probabilities they state, just as physicists do with particles' collisions. But, again, I am sliding off-topic, which'll happen quite a-lot in this chapter of summary (just be aware). So, our body is certainly not our forte; and, yes, it's very-well known that it is our mind, that (literally engineered) energy-sucking-brain of ours, which distinguishes us from the rest of the pack of living creatures. I know, I know, dolphins might have some amazing deeper inner-worlds. I know. But, that is anti-logic, I believe. Understandably, a 'human' is merely the emerging Witness of an event of a mind that's processing data of interactions' events, during the singular event it is alive. In this single event of existence, where elements of entanglements (like memory and meaning) are evident beyond any 'normal' scope of physical-spacetime, a data processing model is manifested as – and not just in-correlation with – Consciousness. So, consciousness is, no-more-no-less, than the sum of information and knowledge and data's entanglements that a single-discrete entity is involved with. Thus, the ultimate reality is most probably – a **Predicting Theory of Everything** (PToE).

Free-Will(ish)

We can't explain any other 'God(s)' than higher levels of consciousness. Well, the highest. Which doesn't mean there's something, anything, human to it. The 'God(s)' may-be more dolphin-like, and less like-the-Sun. Maybe. But, who knows? No maneuver will be of logic, and shall eventually be deemed as dogmatic as any conclusion can ever be. "So, all is known? I have no free will?", you ask once realizing the meaning of the above. Well, it is more like 'Free-Will(ish)', I'd call-it. But if a "yes|no" answer is what you demand here, then I must say: No. Just think about it realistically, practically, and beyond any physics or metaphysics or some 'mambo-shamambo' talk. How much is determined by your pool of genes and your parents' socioeconomic status? If you'd ask our economists, and we saw during the COVID-years that no-one-does, they (including me) would answer that your future salary and health, even your age-of-death, are quite easily 'predicted' (that is educationally-guessed) with just very few variables, including the two I've mentioned. But this type of *Determinism*, of the social and the statistical kind, is not what's of interest.

If we wish to follow this book's scientific (and physics) approach towards such matters, we can utilize Heisenberg's (and Boltzmann's) Uncertainly, and also the clouds-of-electrons' image, in order to obtain some clearer vision of our free-will's borders. Yes, borders again, just as in nature, on all levels. We have realized reality is a game of borders, of bordered events (cells) of interactions, least actions, with no space nor time; that themselves are the building-blocks of all-existence (a body); while all that goes-on within these borders, is always subject-to some basic uncertainly, form the perspective of-all besides the PToE. The concept of a Soul, of an Atman being just another 'cell' of Brahman, and in-complete resemblance with the highest level of consciousness, with God, might be very appealing. But, the reader should already know this is not the case, and we are very-much closer our friends the Dogs and Electrons, than to The Almighty or the Sun. We should understand the boundaries, the awfully strict borders, within which we live (exist). This shouldn't be retrograding our spirits, rather place an acceptance of being human at the hearts of us all.

With relation to the context of free-will and without, again I'll mention and deliver my take on – Artificial Intelligence. It was clearly mocked-at before, this absurdity, but now this term will be explored, just a little-bit. Firstly, it displays the horrible misuse of words by humanity's most professional and educated (and top-earners). We, as a species, still have not agreed-upon what exactly – Intelligence is. However, way-too-many of us are selling, researching, coding and writing-on, some 'Artificial' version of it. Come on you people, my people. Don't you have any shame? Don't get me wrong, I earn more for deploying my Statistical Models, being glorified as AI, yet relay on **Open Source** software that's released by the tech giants. But, that's not the point.

The point is that now 'AI' is not sexy enough, and *Artificial General Intelligence* (AGI) is becoming a legitimate term and reason to ask for a grant or an investment or a salary of half-a-mil (annually).

See, AI is already accepted as 'just' Machines (computers) which run Algorithms that utilize **The Elements of Statistical Learning**[1]. Whoever is telling you otherwise is wrong. Just like drugs, buzzwords require higher dosages. So, they started speaking of – 'AGI'. "This time it is 'General'," they now promise. Why did I want to mention it again, besides having another swing at my colleagues? It was mainly to demystify this term, which is being thrown around as much as – "Quantum"; and also to have a swing at my colleagues from another direction, for their belief that the free-will(ish) of humans is what Statistical Models need to achieve the highest level of – 'General'. On the one hand, humans are all there is, an ultimate reference; on the other, scientists couldn't care less for such creatures.

To summarize, free-will(ish) means a very specific image, and it's not just a movie about an orca in captivity, or some New-Age term. It does not mean we do not have a choice; but, it's merely a semi-conscious decision among several bordered outcomes, within a bordered information-space; and, clearly, around an average. Yes, all existence, free-will including, is of bordered events' facts, including our decisions; that are not only bordered, but are also occurring and mostly found around an average. It's not far-fetched to imagine the human as an entity that is sandwiched

---

[1] Also, it's the name of a textbook by Friedman, Tibshirani, and Hastie (2001), that completely changed my career.

between the averages it sees and interacts with, since some sufficient level of data is optimal, and the borders it is clearly subject-to and active-within. The borders we live-in and are dominated-by, are those of the 'onion-scale-leaf' we sprang from. In humanity's case, of our solar system, probably. We can still be of (some partial) origin from some very distant 'onion-scale-leaf', where the AetherFluid's density (and thus the physics) is somewhat different; but, since the borders we have evolved within are Sun related, they are the ones that determine what we are and how we will operate, more than anything else. This is our upper limit.

From below, we are bordered by the averages we sense – the level of data it was the best fit for us to process in-order to be human (instead of something else entirely, or just not be). But, I feel as-if I wasn't clear, so allow me to be. Most of what we think, or feel, is not up to us. On an on-average level, this is clearly a fact. Just see how political views are evenly distributed among humans, yet so easily predicted with even a mediocre model. We are much closer to the sea-turtles hatching-out of their eggs and 'run' to the ocean, than to the Moon or the Sun (that greatly influence their surroundings). We are so-damn-sure we are so-much-more than a cute-little-turtle (with fins), that will probably be eaten by some bird. But, we're not, and I'm terribly sorry if it offends you, but it's not my problem. So, where's this (biblical) semi-free-will we possess? It's clearly – the thinking of thinking.

Your Model

"Why?" – is the most important question, and also, the one you shall never-ever ask. Why is this the nature of – "why"? No other model is of more importance than the model of yourself. Yet, even those who build models for a living, are hardly giving attention to the building of their own model, the most important one they will ever build. You are just a 'theory', an ensemble of models; entangled to the 'past', see the 'present', and guessing the 'future'; jointly being updated with every piece of data that you, a **Theory of Yourself**, processes. Thus, knowing 'your own model' is essential. Which brings me to the – why(s). Very most certainly, "why?" – is at science's core. No need to explain why. But, above I was pointing at two very specific 'whys'. The first is the 'Why of Existence'. This 'why' is of zero-interest, and my advice is not to waste a second contemplating this problem. I can't say it does not exist, because many do research and preach and tell-of this one 'why', but, there's just no reason why to accept that. Please, tell me nothing of real-nothingness, or anything close to it (such as the 'fractions of a second' the universe was created in). We can also derive some more earthly, humanly, aspect of this fruitless seeking of meaning where it is simply ungraspable. Thinking about, trying to reason, 'why' an accident had occurred, or some horrific loss and even a (real) falling in (true) love – is just pointless. The seeking of the meaning of anything but yourself, is a complete waste of your precious brain-energy. Once, while I was still writing stories, I've called it – "That's Not Thinking".

On the other hand, your internal 'why(s)' is what you should think of, I truly believe. "Why do I want it?", "why did I do this", "what made me feel that" (besides falling in love, which is just another accident). These are just examples of the most important (type-of) 'why(s)'. These are the questions you should ask yourself, almost as much as you can, if you wish to obtain **Your Model**. Again, as I did throughout this short journey of ours, I'll flatten the argument (to a fault). This time with the help of some poker-playing and proprietary-trading. The intersection of these two fields is enormous, though one is based on cards while the other on many numbers around Bids & Asks (prices). In both these professions, if one has no model – one just won't last. It surely doesn't mean all winning models are alike, nor the most commonly used model will be the best fit to every player (either of tables or the markets). But, and it's certain, without (a) *Signal Processing* (model), you are dead in these two lines-of-business; and I think also in modern life, in-general.

What is Learning? What is this process we humans so-appreciate and very proud of ourselves for being just the best-at-it? From a statistical standpoint, it is the finding of significant patterns in the (structured) behavior of information and data. And significant patterns? These are the events that are strongly, significantly, correlated with the precise probabilities of (many and on-average) events. 'Learning' that a fire can hurt your skin does not count as learning, since it is only the acquiring of a *Cosmic Rule*, and thus it is better fitted to be called – "Discovering". Any learning leads to knowledge, the educated assign of probabilities to events that are in the (WF's) known information, and any proper attention leads to such (real) learning. Attention surely deserves some attention, because attention is data, and vice-versa. Attention is also the essence of memory, so I believe. I think that not-remembering an event is the result of an initial lack of attention to it, more than the erasure of something that was stored in a brain (which I think makes no sense). If you wish to know something and become in real-sync with it; if it's the movements of financial markets or the behavior of players around a casino-table, and even with your own true-self; an attention to details, and the signals they're associated with, is required.

Traders and poker players, the professional ones I mean (and not the tens of millions of gamblers who since 2020 transitioned from sports-betting to markets-trading), might be best described as probabilities-machines. Just like quantum-models, more-or-less. First and foremost, such experts are measured by their WFs, their experience and reasonable imagination. As in life, the concrete foundation of such experts' model is – information. Knowing life's potentials will keep you unsurprised by even the most extreme of events, and will lower the possibility of life catching you with your pants down. It's more than half the business, and no model, no system of betting (or living), can even be imagined without it. Both traders and poker-players should always be aware of what can happen (the borders of events), much more than being on-the-money with the probabilities they educationally-guess for those borders' happening (averages of events). Don't forget, any existence is subject-to randomicity and higher-levels of Dark Variables, that together are stronger than any human-model. This is why, not knowing the borders of you, and the ones you are operating within, is what'll get you bankrupted, at best, and probably hurt or dead if you are playing a game you should not be playing. At the heart of any method of trading or poker-playing, and surely of living, lies some model of signal processing. And signals, both for traders and other educated-bettors, and for plain-humans, are the variables in the models that should help the trader and 'pokerer', and you, getting-by.

What you should aim for, and I believe this is what it means to be an adult, is the creation of your personal, internal and cosmic, hinting-model. Let me explain. Data, what your mind consciously and unconsciously reacts to, is of three origins. One more abstract and two very mundane. The latter two are quite obvious, since our brains process data of the body and the senses, only. "Know thyself," should've been the first commandment, since knowing and predicting yourself is the first step of obtaining the (very) limited (to begin with) free-will that you, we, possess. The question you should be asking yourself first thing when you open your eyes and realize you did not die in your sleep, is: "Should I get out of bed today?" It truly does not matter if you must, because some days are better to be spent with the minimum interactions possible, or even, avoid them all whatsoever. If you'll be able to derive such a model, receiving as input your first feelings and thoughts, and predicting the probability that today is going to be a bad-day, you are half-way-there. The other half, of having a model to deal with the day's remainder, is a different story.

Experience is of the essence, because experience is the accumulation of experiments and observation, and their entanglements; that are nothing but data-points which sprang from the information of space by an event of interaction, including the event of you giving your attention. This buildup must result with a greater knowledge of yourself, knowing your reactions and typical turn-of-events (which is the reality you experience), if attention and self-analysis are applied with care. I know how it sounds, this call for such constant self-observing and experimenting. Yes, it may seem like **Elements of Schizophrenia** are of-virtue, if and when are under-very-strict-control. Which smoothly thrusts me to cosmic hinting, and the entire issue of our relationship with the universe – with life as-a-whole.

It is quite tempting to see 'signs' all around. You don't have to be crazy or insane to experience that. Ask any trader or poker-player (especially the trader); they see signs all the time and everywhere, and that's the way it should be. The experiencing of life and the universe is also the experiencing of yourself, and external-hinting should be modeled and become systematic just as internal-hinting, that some call – "Feelings". Any model of yourself must include the following. The objective data coming to your mind through the senses, the internal data that your body and mind produce, and those cosmic 'signs' that you may become aware-of during the course of your life. Your Model must incorporate all the noise and signals, and know to distinguish between the two, that are manifested in those data-points you both consume and internally-produce. The more you observe yourself, the more attention you give to the giving of attention (and not just the mere passive-act of reacting to events); the more you will grow and become-identified, mainly, with your true-self; and, less you'll get-attached to what others think-of, and see-in, you.

Data Consumption

If I have named the integrated metaphysics offered here – of Vedanta's philosophy with physics and consciousness' facts – VeData, I probably assign data some fundamentality and importance. I'm also pretty sure thinking in such a manner and logics, regarding consciousness and mass, is helpful (and maybe even right). People may say I suggest it only due to the facts that I'm a data scientist who's completely engulfed in the industry of models and analytics, but I promise this is not the case. Beyond any other motive, I had wished to focus on Data since our physicists got entirely blundered with it, as much as they are with anything else, and completely (and repeatedly) confuse it with Information.

In her book, "The Science of Can and Can't: A Physicist's Journey Through the Land of Counterfactuals" (2021), which deals mainly with information (yet somehow not with data) within the framework of *Quantum Computing*, Chiara Marletto does not mention data even once. She does mention 'artificial intelligence' three times, which you know what I think about; and also some grand *Universal Constructor*, which is quite close to the Gods' Eye I've illustrated here. She also mentions the word "model" six times, in the same page, and that's it. How can you tackle physics without dealing with models? Yet she speaks much of "theories", effortlessly jumping between describing both models and theories. It is hard to believe, but even (physics) information 'experts', who believe there are 'classical' and 'quantum' types-of-it (again thinking in-contrasts of 'big' and 'small'); just have-no-idea how to distinguish between data and information, nor theories and models, nor what is 'intelligence'. They don't realize they keep-on shoveling blunders in physics.

This is how physicists (mistakenly) think of data; and pretty Chiara is a fine example of their minds, this I'm sure-of. But you, you know much better, I'm pretty sure. Data is data – hard to argue with that. Data is the same for the waves' interactions and those clips you watch on TikTok. Let me quickly remind. The (Space's) AetherFluid and Information are of the same nature, a Wave Function is their (approximation) model; while, Data (Facts of Interactions' Events) and Mass (probably made-of and carried-by Light) are completely identical. Elementary Particles, as 'seen' in colliders, (and that is certain) are just short-lived data-points. And Points? These are the Knots of AetherFluid's (spherical) waves, that become so-dense and Entangled – that Time is (relatively) created (for every entity other than Gods' Eye). By the way, it can also be called – God's Eyes, and some may say they imagine Blackholes as their manifestation. This is the story I tell myself; the only story, considering the very few facts that physicists and scientists do-agree-upon, that makes sense (to me). In such a metaphysics, data is of the essence for all besides the one Gods' Eye (or God's Eye, or even God's Eyes; but certainly not the Gods' Eyes, which simply makes no sense; just like the many-parallel-worlds of the Multiverse). And since we are living in the Age of Big Data, which started at 2012, data needs more clarification.

Let's think about it some. I'll start with a question. Are the 2021's humans consuming more data than the 1930's? A simple question, which should be easy to answer using the visions, the 'hallucinations' and structures, that were constructed here, I believe. In the chapter on Seeing and Psilocybin, and the inter-and-outer Entanglement of photons into images of (conscious) meaning; it was stated that our human (data) consume-ability and interpretability are hard-coded into our being, and that psychedelics are an example of tools for some (momentarily) going-beyond the optimal human mundanity. This clearly means that none-has changes with the human-physic since the 1930s. What has greatly changed, greatly evolved, are all the elements of existence associated with consciousness. Consciousness, as a 'Mother-Theory' that incorporates all the models you build (that you are) in-order to survive, evolves with every piece of data that gets consumed and entangled; which leads to growth in information and knowledge, that way or the other. But that – "that way or the other" – is exactly, precisely, what needs to be discussed now.

The ancient humans, I mean nature's humans and not the city-mice, had built their religion, their consciousness, solely around nature's objects. That was what they saw and were unable to go beyond. I think we are experiencing a similar phenomenon in the Age of Big Data. Think of those fucking Nazis and that sadistically-inspirational Hitler. They were the 'right' people at the 'right' time; just like Michael Jeffery Jordan, but of a different kind – of the bat-shit crazy one. TV was introduced to Germany in the mid-1930s, and if you think it had nothing to do with that nation being brainwashed into the biggest cult ever, then you are either asleep with your eyes open or of-slim understanding in humans. Try to imagine the impact the first TV, the first screen, had on humans' consciousness and the spreading of ideas. Ah, in-fact, it is not that complicated to imagine, because we are all currently witnessing the impact that **Online-Porn** had on humanity – on its sexuality. Do you think the way you are (ab)using that mini-machine that you carry in your pocket has no impact on your-consciousness, on you? You are just your own Data Consumption.

I can only hope you realize what this means, and that this (preaching) paragraph is completely not-necessary; but, I still feel it must be said explicitly. Social Relativity was already mentioned, and I think it's the most dominant form of human consciousness, the only consciousness possibly imagined, that is identified throughout the modern world. Well, there is nothing new here, there is really nothing new under the sun, only a change in-volume and thus in-impact on the personal and the collective consciousness. No-doubt, humans have always cared-for what other humans think-of them. Only now, they have access to statistics of how many 'Like' (or not) what they, or their favorite figure, think, or how they look; and, those numbers are changing by the second, like a slot-machine. People used to process incoming data thorough a grid of a tribal-god, and how they had believed it views them. Many humans still think this way, mainly the Monotheistic ones, and I am not sure this old-grid (or theory) is less-healthy than that of Social Relativity.

An Age of Social Science

I was undecided whether I should finish this while I go-down swigging, or begging. With the current state of civilizations, mostly those occupying the developed world, being on the verge of a Cultural Civil War, I thought it's better I'll follow the latter. Nevertheless, I do believe that some-kind-of-warning is still very-much-in-place. Besides some begging and warning, some macro-guidance for science, beyond any personal guidance, should also be presented here. I also plan to conclude this book with a (hazing) statement, that (statistically) you won't like; but, you should already know, I-care-not. 'Feelings' – for me and other Vedantists – are no-more than data-points we witness. Yes, feelings are just the internal signs, quite of-significance, you should be incorporating into Your Model.

If there's one thing I hate, is when people speak 'for the people'. Having said that, you know what comes now. Dear scientist-reader, or scientist-to-be, I'm warning thee. Trust me, 'the people' have had enough. "Who are those people and what have they had enough of?", you ask. Science is no longer a realm of scientists-only, in the same manner that Social Healthcare is no longer a field for medical doctors and epidemiologists alone. We saw how the entire world's physical and economic wellbeing were (so-very-poorly) managed; while the mental side of the human existence, and our democratic foundations, were neglected at-best or run-over at worst. Maybe this is the evolution, the higher level of awareness to (real) reality, that was hastened by the realizations of the COVID-years.

This also means that all those billions of $s or €s or £s that were and still are spent on experiments that won't be directly, the-day-after, beneficial to humanity and society, are simply no-longer-acceptable. We are at times of emergency, I believe; both physically and mentally, I'm pretty certain. Therefore, I think that just for now, and until further notice, science must follow this very-simple-rule. Will it (whatever it is that you wish to research) directly advance equal opportunities, or won't it. Easy, isn't it? It does not mean that all scientists should follow the same lead, nor that none should advance other fronts of their field. But, on-average, just as they love it, scientists, through their institutions of science, should advance social benefits. Leaders of the West are preoccupied with alarming their tax-payers of outside physical threats (when they don't inflate the deadliness of viruses); while it's the pandemic of inequality, which grew and deepened under COVID, that must frighten them most. I think that **Wealth Redistribution** is closer than most would believe.

And absolutely final last words. You must know that Bitcoin's worth is jack-shit-zero. I understand, sometimes some do find another stupid-motherfucker to buy it from them for a higher amount of dollars (which is what they are really after); but (and because you are reading these words you must really be valuing my thoughts), believe me when I say this is a classic case of a commodity-bubble. Unlike Psychedelics and Physics, that you should be investing both your attention and money in, Bitcoin is a (real) hallucination.